# Urban Transport and Hybrid Vehicles

edited by
**Seref Soylu**

**Urban Transport and Hybrid Vehicles**
Edited by Seref Soylu

**Published by InTech**
Janeza Trdine 9, 51000 Rijeka, Croatia



**Publishing Process Manager** Jelena Marusic

**Technical Editor** Teodora Smiljanic

**Cover Designer** InTech Design Team

Additional hard copies can be obtained from orders@intechopen.com

Urban Transport and Hybrid Vehicles, Edited by Seref Soylu
p. cm.
ISBN 978-953-307-100-8

For EU product safety concerns:
IN TECH d.o.o., Prolaz Marije Krucifikse Kozulić 3, 51000 Rijeka, Croatia,
info@intechopen.com or visit our website at intechopen.com.

# Meet the editor

Dr. Seref Soylu received his Ph.D degree in Mechanical Engineering from Iowa State University in 2001. His Ph.D. research work mostly focused on internal combustion engines and thermodynamic engine modeling that were supported by John Deere Product Engineering Center. Through end of his Ph.D. study, he also worked for Caterpillar Inc. as an analytical engineer in their technical center in Peoria, IL for the development of the advanced diesel engines. After receiving his Ph.D. degree in 2001, Dr. Soylu joined to Sakarya University to perform research and teach thermal science courses. His research interests centered on the development of environmentally friendly road transport vehicles including hybrid electrical vehicles. Dr. Soylu also worked as a visiting scientist in the Joint Research Center of European Commission for a year from November 2004 to November 2005 to provide technical and scientific support to the policymakers of the Commission for their legislative works. His work in the Joint Research Center focused on fuels for internal combustion engines, exhaust emission measurement techniques for zero emission vehicles, portable emission measurement systems, and small engines. Dr. Soylu has also worked as independent expert for European Commission in the field of Energy and Transport to evaluate and review Framework Programme (6 & 7th) projects on behalf of the Commission. Dr. Soylu is currently an Associate Professor of Sakarya University. He is teaching undergraduate and graduate level thermal science courses and leading a research project entitled "Measurement and Modelling of Hybrid City Bus Real-World Emissions" funded by Turkish Ministry of Industry and Trade and TEMSA R&D.

# Contents

# Preface

Urban transport system is crucial for economic and social development of our society as it meets challenging mobility requirements of urban agglomerations. The requirements are challenging because road transport, which is the most problematic transport mode, is dominating the urban transport activities. The main problems with the road transport are that the infrastructure is not generally sufficient for whole traffic conditions and the vehicles are powered by internal combustion engines that require fossil fuels to burn and emit pollutant emissions during their normal operations.

Internal combustion engines have been used successfully for almost a century as power source of road transport vehicles, but in the same period vehicle ownership increased to a level that finite fossil fuel resources and environmental and health impact of the emissions have become great concern for a few decades. There are several regulations to be issued to promote alternative fuels, however, fuel demand of urban transport vehicles reached to a level that alternative fuel production levels are far from meeting the demand. On the other hand, the exhaust emissions are growing concern for especially urban transport vehicles because these vehicles operate where people normally lives and emit harmful emissions. In the streets of urban agglomerations, there is not much time and distance for the emissions to be diluted to a harmless concentration before the emissions are inhaled by human receptors.

This book is the result of valuable contributions from many researchers who work on both technical and nontechnical sides of the field to be remedy for typical road transport problems. Many research results are merged together to make this book a guide for industry, academia and policy makers. I hope you will get maximum benefit from this book to take the urban transport system to a sustainable level. As the editor of this book, I would like to express my gratitude to the chapter authors for submitting such valuable works that most of them already published or presented in scientific journals and conferences.

The chapters of the book are designed in a logical order. It is started with the examination of typical problems of conventional city busses. Then, an overview of the current state of the art traffic flow models is presented. It is well known that traffic flow models are vital requirement for advanced traffic management systems. Finally, hybrid electrical vehicles as an alternative to the conventional vehicles are examined in details. Hybrid electric vehicles are expected to be remedy for the minimization of noise, fuel consumption, and pollutant emissions of typical conventional vehicles.

August 12, 2010

Editor

**Francisco J. Gallegos-Funes**
*Sakarya University*
*Department of Environmental Engineering*
*Sakarya- TURKEY*

# 1

# Examination of an Urban City Bus Operating Conditions and Emissions

Seref Soylu, Ayda Bal, Hülya Semercioglu and Eyup Fatih Ay
*Sakarya University*
*Turkey*

## 1. Introduction

City busses are main vehicles for public transport to meet travel demand of the society. They operate where urban population is very dense and release such emissions as particulate matters (PM), nitrogen oxides (NOx), carbon monoxides (CO), and hydrocarbons (HC) where the population lives (Soylu et al., 2009, Gumrukcuoglu et al., 2008, Soylu, 2007, WHO, 2003). Unfortunately, concentrations of the released emissions are generally high enough to damage human health and there is no enough time for the emissions to be diluted in the air to harmless concentrations before they are inhaled by human receptors. It is well known from the literature that exposure to even relatively low concentrations of vehicle emissions exacerbates or provokes many diseases (WHO, 2005a, WHO, 2005b). Adverse health effects of the emissions have been known for many decades and in order to prevent these effects many strict legislations, which reduces the limits more than 90% over four decades, for vehicle emissions have been issued. However, urban populations in many developed countries are still suffering from urban transport sourced emissions (Duclaux, 2002, Colvile et al.,2001, Frey et al., 2009, Erlandsson et al., 2008). One of the important reasons for this is that the engine certification test cycles don't represent the real world in-use operation of the vehicles and, hence, quantity of vehicle emissions to be released in the urban streets has not been reduced in parallel with the stringent emission legislations (Cocker et al., 2004, Lents et al., 2007).
City buses generally use diesel engines as power source and emit carbon dioxide ($CO_2$), water vapor ($H_2O$), and nitrogen ($N_2$) to the ambient air as the main products of engine combustion. The quantity of $CO_2$, which is the major greenhouse gas (GHG), is proportional to the fuel used in transport activities and it is inevitable combustion product, however, it is not necessary to emit CO, HC, NOx, and PM, which are generally called local pollutants since they are more harmful where they are emitted. The emissions of the local pollutants from city busses depend strongly on engine combustion technology, exhaust after-treatment devices, fuel quality, vehicle aging, and operating conditions (Cocker et al., 2004, Lents et al., 2007, Regulation, 2009). Especially the bus operating conditions may have significant effects on the emissions. Depending on city traffic and road conditions which involves many short trips with frequent accelerations, decelerations, low rush hour speeds and various road grades, the emissions may change with an order of magnitude (Cocker et al., 2004).
Urban transport operating conditions and especially the city bus operating conditions are quite specific for a particular city and, hence, the well known certification test cycles cannot represent accurately these conditions all over the world. For this reason EURO VI regulation

for heavy duty vehicles requires application of portable emissions measurement systems (PEMS) for verifying the real world in-use and off-cycle emissions (Johnson et al., 2009). PEMS are remedy for the real world emission measurement as they can be installed quickly to the vehicle and measure in-use emissions. There are varieties of PEMS available on the market to measure both gaseous and PM emissions (Lents et al., 2007, Durbin et al., 2007). Some of them have minor differences in measurement theory and technique that may require standardization of PEMS themselves in the future (Younglove et al., 2005). With PEMS and their auxiliary systems it is possible to examine the impacts of road and traffic conditions on the performance of the vehicle, the engine, the after-treatment system, and the exhaust emissions. Therefore, it is possible to optimize a vehicle in terms of fuel consumption and emissions for a specific route by using PEMS. Besides real world emission performance of advanced technology vehicles such as hybrid vehicles should be tested by using PEMS since they have an electrical drive unit, also. It is crucial for the automotive industry as manufacturers and the municipalities as end-users to choose the most suitable vehicle for the specific transport route to minimize the capital cost, the fuel consumption, and the emissions. In this sense the PEMS are one of the most useful equipments to meet this need.

PEMS are also very useful in quantification of transport sourced emissions. Since the emissions factors can be determined under real world conditions, determination of the effects of transport sourced emission on the global air quality as required by EURO VI can be much more accurate (Johnson et al., 2009). There have been various research work published in the literature to examine real-world emission of the vehicles to develop emission factors and prepare inventory for a vehicle class (Durbin et al., 2007, Younglove et al., 2005, Durbin et al., 2008, CFR Part 86, 2008). The US EPA and EC JRC have made significant effort for development of proper methodologies for the real-world in-use emission measurement as a part of transport emission regulations (Soylu et al., 2009, Southwest Research Institute Report, 2008). The US EPA together with CARB, SENSORS Inc. and Caterpillar Inc. initiated a programme to develop a mobile emissions laboratory (MEL), which is compliant with code of federal registration (CFR) 1065, to compare and validate accuracy of PEMS under different in-use driving conditions (Lents et al., 2007, Durbin et al., 2007, CFR Part 86, 2008, Southwest Research Institute Report, 2008). At the conclusion of this programme the final measurement allowance value for NOx emissions determined to be as high as 4.5%.

Swedish Road Administration was also carried out a PEMS programme to test in use performance of heavy-duty vehicles under real world conditions (Cocker et al., 2004). In this programme three city busses (which are Euro IV/V level) were equipped with PEMS and tested on a reference route which comprises urban, rural and highway driving with an average speed of approximately 60 km/h and on an actual bus route. It was observed that on the actual bus route, high level transient operation of the city busses have significant adverse effects on the performance of exhaust after-treatment system. Especially the NOx emissions were increased almost up to an order of magnitude when compared to that of the reference route.

In Turkey, a similar research project has been introduced by Sakarya University with support of Turkish Ministry of Industry and TEMSA R&D, to quantify impacts of hybridization on city bus emissions and fuel consumption under real world in-use conditions (Soylu, 2009). In order to quantify the impacts, a two-phase test programme was prepared to measure real world in-use emissions and fuel consumptions of both conventional and hybrid city busses on a specific bus route in Sakarya city center. In the present work the results from the first phase of the test programme that involves examination of the effects of road and traffic conditions on conventional bus operating characteristics and the response of bus engine and selective catalytic reduction (SCR) system to these characteristics in terms of NOx emission were presented.

## 2. Experimental procedures

In this work all the tests were carried out on the "university route" of Sakarya Municipality city busses. The route includes a round trip between Sakarya University and the city center which represents typical Sakarya urban driving. It is 22 km long and involves 48 bus stops and the altitude changes from 30 m to 220 m. The travel time is approximately 60 minutes. During the test the vehicle was driven on the route with and without its SCR system and data for vehicle speed and location, engine operation characteristics, exhaust flow-rate, exhaust emissions and environmental conditions were sampled second by second.
The test measurements were made by using a SEMTECH DS from SENSORS inc. This system includes a flame ionization detector (FID) for total HC measurement, a non-dispersive infrared (NDIR) sensor for CO and CO2 measurement, a non-dispersive ultra violet (NDUV) sensor for NO and NO2 measurement. The NDUV measurement is different from standard chemiluminescence measurement that is used for the reference method. Exhaust mass flow-rate was measured by using SEMTECH EFM which operates based on pitot tube technology. Before the test the PEMS was warmed up according to recommended operating conditions and then, zero, span and audit calibrations were completed to ensure the accuracy.
The test vehicle was TEMSA AVENUE which is a 12 m long city bus. It is powered with a 6.7 liter engine CUMMINS ISB EURO 4 (model year 2005) engine which produces 250 HP at 2500 rpm. The engine was certified to 3.5 g/kWh NOx standard. Over the test the ambient temperature varied from 25 to 30 °C and the relative humidity varied from 30 to 45 % which are typical for summer time.

## 3. Results

Figure 1 indicates a speed profile for the specific test route, which is a real world city bus speed profile. Highly transient behavior of the route can be seen clearly from the figure. The vehicle speed reaches to as high as 65 km/h but there are many stops with corresponding decelerations and accelerations because of the road traffic and bus stops. It is well known from literature that vehicle speed profile has a strong impact on operating characteristics of the vehicle engine. Figure 2 indicates the effects of the real world drive characteristics on the engine map in terms of the number of occurrences (frequency) corresponding to engine loads and speeds. As can be seen from the figure, under the real world drive conditions the engine operates most of the time at loads less than 60% and low speed conditions. Figures 3 and 4 indicate engine load-speed map for European Transient Cycle (ETC) that is current regulatory test cycle for heavy duty engines and World Harmonized Transient Cycle (WHTC) which is expected to be valid certification cycle for heavy-duty engines with EURO 6 regulation. As can be seen from the figures in these cycles the engine operates mostly at speeds between 1200 and 1500 rpm and at loads less than 50% except for ETC which also operates frequently at loads higher than 90% as well. From these maps it is easy to see that the certification test cycles don't correlate very well with the real world city bus driving conditions although city bus engines are still certified according to ETC. WHTC operates more on the low speed conditions but, the frequencies on the load-speed map are still significantly different from that of the real world city bus driving conditions. It is therefore questionable if the certification cycle emissions limits ever be realized in the real world driving conditions. A certification test cycle should be able represent real world operating conditions of an engine, however there are different type of use of these engines such as highway trucks and busses, delivery trucks, and city busses. Engines of these different vehicle classes will certainly operate with different frequencies on the engine load-speed map although they all are certified with the same test cycle. Therefore, off-cycle operation of these engines and corresponding emissions can be significantly higher than that of the certification test cycle. In order to minimize these off-cycle emissions, EPA introduced Not To Exceed (NTE) regulations but there is almost no way to completely control engine emissions when considering sophistications at engines and their after-treatment systems.

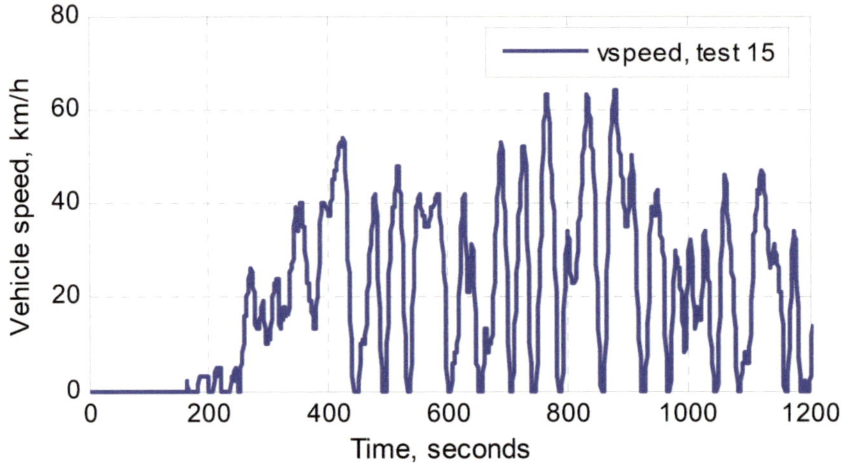


Fig. 1. Variation of the vehicle speed on the specific bus route

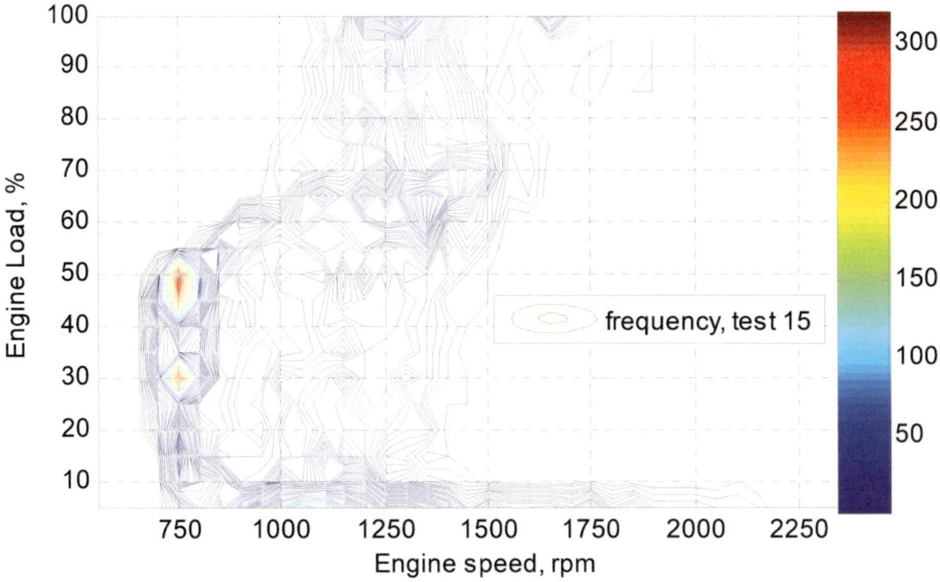


Fig. 2. Engine load-speed map for real world driving.

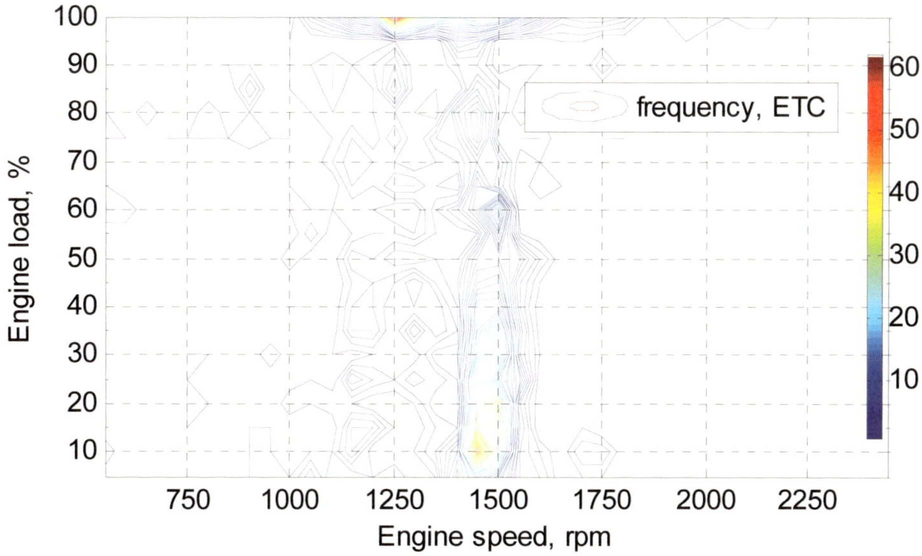


Fig. 3. Engine load-speed map for ETC

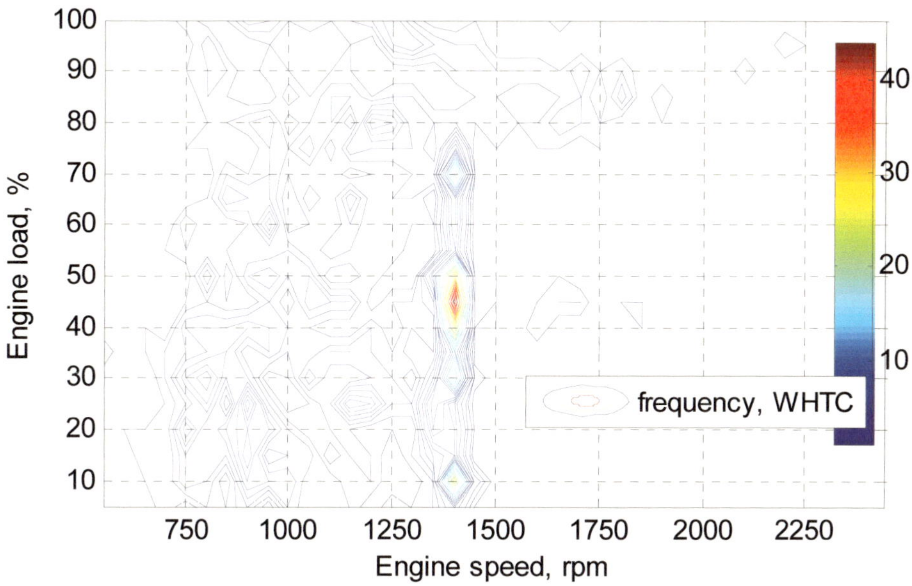


Fig. 4. Engine load-speed map for WHTC (to be valid with EURO 6)

As mentioned earlier, vehicle emissions are strongly dependent on engine operating conditions. Especially NOx emission formation is almost directly increase with the engine load since the formation is dependent strongly on combustion temperature and the available oxygen. As can be seen from Figure 5a and 5b, there are two common strategies to minimize NOx emissions. First approach is the exhaust gas recirculation (EGR) which minimizes NOx formation in the engine by lowering the oxygen concentration and maximum combustion

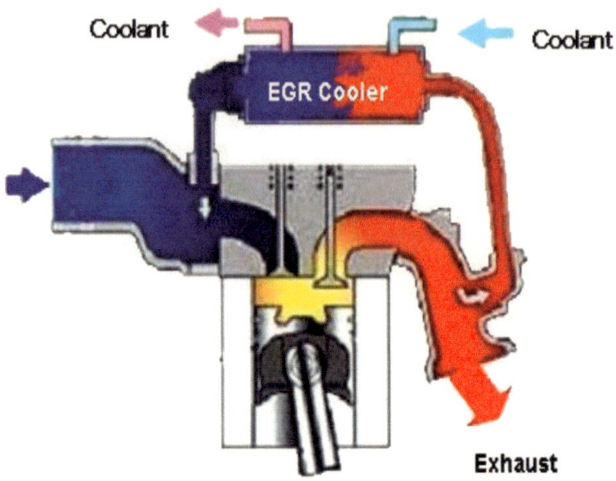


Fig. 5a. EGR strategy to minimize NOx emissions (TRAD, 2010).

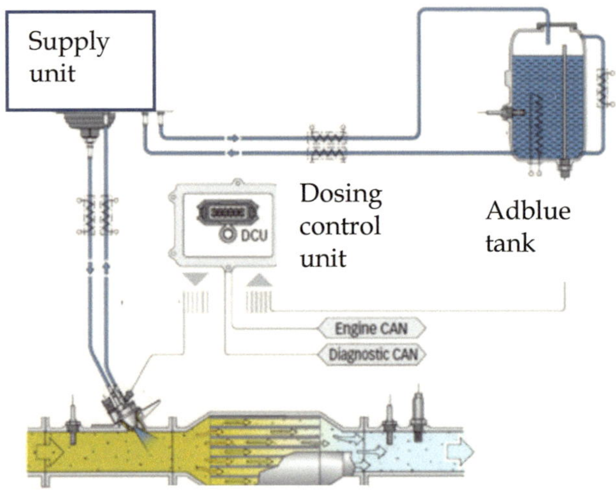


Fig. 5b. SCR strategy to minimize NOx emissions (VALTRA, 2010).

temperature. However, this approach may have an adverse effect on PM emissions and the fuel consumptions. The second approach is selective catalytic reduction (SCR) of NOx emissions which requires a catalyst and aqueous urea solution on the vehicle. Once the urea solution is injected to hot exhaust gases, $NH_3$ is formed from the injected urea by means of hydrolysis. Then NOx emissions are minimized in the catalyst after its chemical reaction with the $NH_3$ to form nitrogen and water. SCR of NOx emissions doesn't have any adverse effect on fuel consumption and PM emission since the engines can be tuned for maximum thermal efficiency. However, there are a few other concerns with this approach. First of all, freezing temperature of the urea solution is quite high, which is -11 °C, and this may cause technical problems during winter time (Koebl, 2000). The second is that high values of $NH_3$ slip may result if the stoichiometric ratio of NOx to $NH_3$ cannot be provided. The main difficulty with this stoichiometry is that engine load and speed are not steady and hence the concentration of NOx emissions in the exhaust. Therefore establishing the stoichiometry is mostly dependent on precision of the urea injection control system. The third concern with SCR system is that it is temperature dependent. In order to reduce the NOx emissions efficiently, the exhaust gas temperature in the SCR system must be higher than 250 °C. Therefore, the reduction of NOx with SCR system is highly dependent on engine operating conditions and, hence, vehicle driving conditions.

Figure 6 indicates exhaust gas temperature distribution over the engine load-speed map for the specific bus route. As can be seen from the figure, the temperature is generally around 250 °C over the entire map although there is a tendency that the temperature rises up through higher loads and speeds. Figure 7 indicates NOx emissions in grams per second to be released over the engine load-speed map corresponding to the same bus route. As can be seen from the figure, the NOx emissions increase with the higher speeds and loads because of the higher combustion temperatures and higher exhaust flow-rates associated with it.

Figure 8 indicates the distribution of the total NOx emissions to be released in grams over entire trip. As can be seen from the figure during this trip most of the NOx emissions were released at a location on the map where load is about to 45% and speed is about to 750 rpm, which are quite low. This seems to be surprising, because as was given in Figure 7, the NOx emissions in terms of grams per second was increasing with higher loads and speeds. However, as was given in Figure 2 during the entire trip the engine operating frequency at 45% load and 750 rpm speed was shown about 300 that is one of the most visited locations. For this reason the total NOx emissions to be released during entire trip is highest at this location. This location probably corresponds to the city bus idling and first acceleration, which is visited during every bus stop, for this reason this location is visited most during the entire trip.

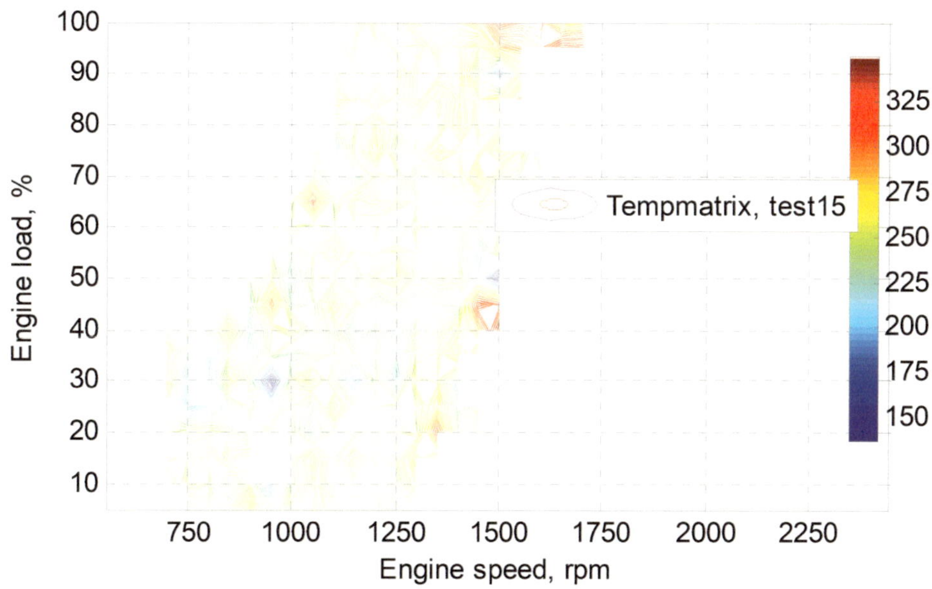


Fig. 6. Exhaust gas temperature distribution (°C) over engine speed - load map (Temperature measured from exhaust flow meter).

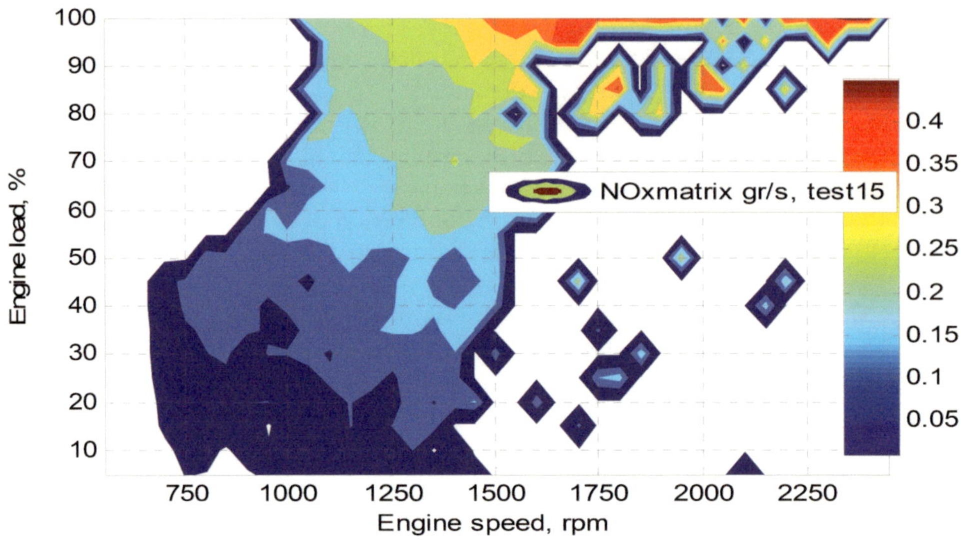


Fig. 7. NOx emissions to be released over engine load-speed map without SCR system

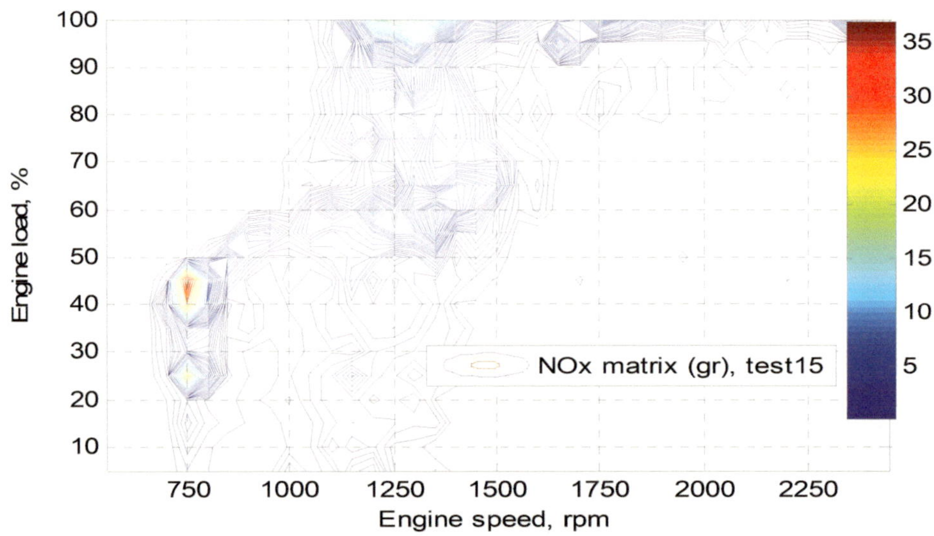


Fig. 8. Cycle NOx emissions to be released over engine load-speed map without SCR system

Figure 9 indicates another trip on the same route while SCR system of the city bus is active. As can be seen from the figure, NOx emissions to be released over the entire map are much less than that of the trip without SCR system, which is Figure 7. NOx emissions are efficiently reduced exceptions for a few hot spots. Besides, there is no step wise increase in NOx emissions with increasing load and speed exception for loads higher than 90%. At this load and speed the exhaust mass flow-rate is probably so high that the NOx emissions in grams per second are still significant. Figure 10 indicates the total NOx emissions to be released for the entire trip while SCR system is active. As can be seen from the figure the

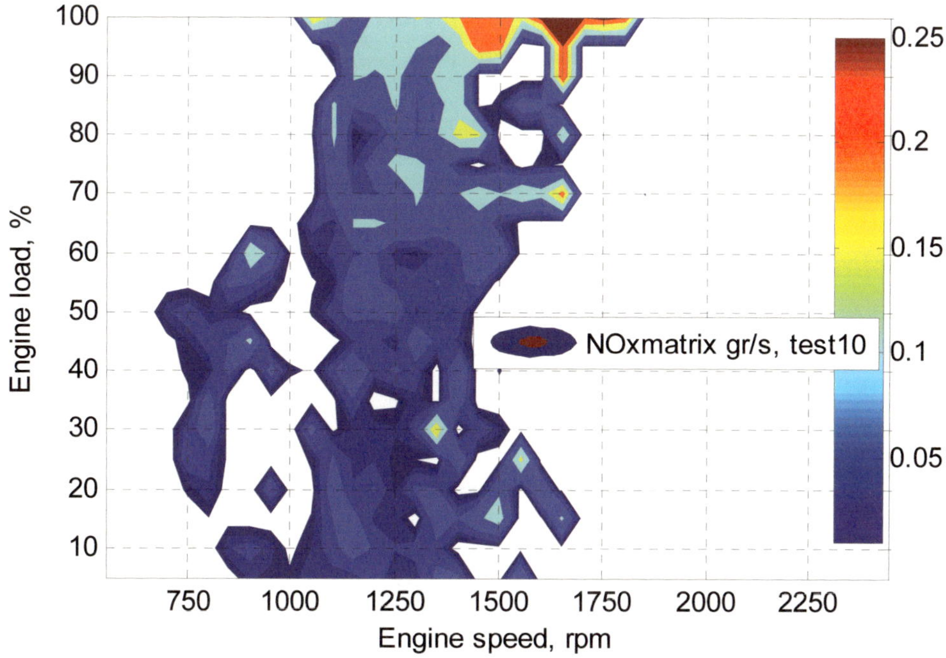


Fig. 9. NOx emissions to be released over engine load-speed map with SCR system.

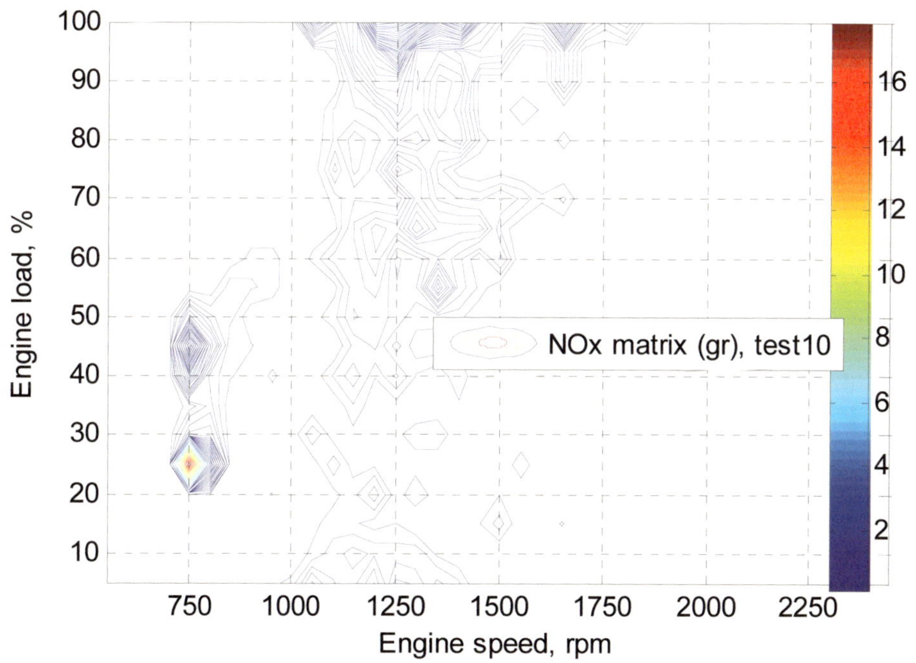


Fig. 10. Cycle NOx emissions to be released over engine load-speed map with SCR system.

most of the NOx emissions are released at 25% load and 750 rpm speed. This location is the idling location and as was given in Figure 2, it is one of the most visited locations during the entire trip on this route. For this reason, the total NOx emissions to be released during the entire trip are highest at this location.
These figures indicates that city busses that operate in the bus routes are generally operating at low loads and speeds and most of the trip emissions are corresponding to these locations on the map.

## 4. Conclusions

It was observed that real world driving characteristics of the city bus don't correlate well with that of the ETC, which is the certification cycle for the city bus engines. For this reason it is not realistic to expect emission levels lower than corresponding EURO standards from the city bus during real world driving. City bus NOx emissions in terms of grams per second are increasing with engine load and speed but the total NOx emissions for entire trip are the maximum at the locations where engine visited most, which are low speed and load conditions. Therefore, while optimizing the city buses for the best fuel economy and emissions, the real world drive characteristics of the city bus and the corresponding operating conditions of the engine and after-treatment system must be considered.

## 5. Acknowledgment

TEMSA R&D and Turkish Ministry of Industry and Trade are acknowledged for their contribution to this work.

## 6. References

[1] Soylu, S.; Gumrukcuoglu, M; Senturk, G.; Bal, A.; Aydınlar, B. (2009). Examination Urban Transport Activities And Emissions, Adapazarı Case Study, the *17th transport and air Pollution Symposium and the third Environment and Transport Symposium*, June 2-4, 2009 Toulouse, France

[2] Gümrükçüoğlu M.; Soylu S. (2008). Monitoring of urban transport sourced pollutants by using GIS, *5th International Conference on Geographic Information Systems*, July 2-5, 2008, Istanbul-Turkey

[3] Soylu, S. (2007). Estimation of Turkish Road Transport Emissions, *Energy Policy* 35 (2007) 4088–4094, April 2007.

[4] WHO, 2003. Health aspects of air pollution with particulate matter, ozone, and nitrogen dioxide. Report on a WHO Working Group, Germany, 2003. www.who.int/en

[5] WHO, 2005a. Air quality guidelines for Europe; second edition Copenhagen. WHO Regional Office for Europe, WHO Regional Publications. European series; No 91. www.euro.who.int/air/activities/20050223_4

[6] WHO, 2005b. Health effects of transport-related air pollution. WHO Regional Office for Europe Scherfigsvej 8, DK-2100 Copenhagen , Denmark

[7] Duclaux, O.(2002). 3D-air quality model evaluation using the Lidar technique. *AtmosphericEnvironment, 2002*- 36(32), 5081-5095.

[8] Colvile, R.N., Hutchinson, E.J., Mindell, J.S., Warren, R.F. (2001). The transport sector as a source of air pollution. *Atmospheric Environment 2001*-35, 1537–1565.

[9] Frey, H. C.; Zhai, H. & Rouphail, N. M. (2009). Regional On-Road Vehicle Running Emissions Modeling and Evaluation for Conventional and Alternative Vehicle Technologies. *Environmental Science & Technology 2009- 43,* 8449–8455

[10] Erlandsson, L.; Almen, J, & Johansson, H. (2008). Measurement of emissions from heavy duty vehicles meeting Euro IV/V emission levels by using on-board measurement in real life operation. *16th International Symposium 'Transport and Air Pollution*

[11] Cocker III, D. R.; Sandip, D. S.; Johnson, K.; Miller, J . W. & Norbeck, J. M. (2004). Development and Application of a Mobile Laboratory for Measuring Emissions from Diesel Engines. 1. Regulated Gaseous Emissions. *Environmental Science &Technology 2004-* 38, 2182-2189

[12] Lents, J.; Unal, A.; Mangir, N.; Osses, M.; Tolvett, S. & Yunusoglu, O. (2007). A study of the emissions from diesel vehicles operating in Istanbul, Turkey

[13] Regulation, (2009). EC Regulation No 595/2009 of the European Parliament and of the Council *18 June 2009*

[14] Johnson , K.C.; Durbin, T.D.; Cocker III, D.R.; Miller, W.J.; Bishnu, D.K.; Maldonado, H; Moynahan, N.; Ensfield, C. & Laroo, C.A. (2009). On-road comparison of a portable emission measurement system with a mobile reference laboratory for a heavy-duty diesel vehicle. *Atmospheric Environment* 43 (2009) 2877–2883

[15] Durbin, T.D. & Johnson, K. (2007). Evaluation and Comparison of Portable Emissions Measurement Systems and Federal Reference Methods for Emissions from a Back-Up Generator and a Diesel Truck Operated on a Chassis Dynamometer. *Environmental Science & Technology 2007*-41, 6199-6204

[16] Younglove, T.; Scora, G. & Barth, M. (2005). Designing on-road vehicle test programs for effective vehicle emission model development. *Revised for Transportation Research Record*

[17] Durbin, T.D.; Johnson, K.; Miller, J.W.; Maldonado, H. & Chernich, D. (2008). Emissions from heavy-duty vehicles under actual on-road driving conditions. *Atmospheric Environment* 42 (2008) 4812–4821

[18] CFR Part 86, (2008). In-Use Testing for Heavy-Duty Diesel Engines and Vehicles; Emission Measurement Accuracy Margins for Portable Emission Measurement Systems and Program Revisions, EPA, 40 CFR Part 86, Federal Register / Vol. 73, No. 50 / Thursday, March 13, 2008 / Rules and Regulations

[19] Southwest Research Institute Report, (2008). Determination of PEMS Measurement Allowances for Gaseous Emissions Regulated Under the Heavy-Duty Diesel Engine In-Use Testing Program, Assessment and Standards Division Office of Transportation and Air Quality U.S. Environmental Protection Agency, Prepared for EPA by Southwest Research Institute EPA Contract No. EP-C-05-018, February 2008

[20] Soylu, S. (2009). Measurement and Modelling of Hybrid City Bus Emissions under Real world Conditions, SANTEZ Project 6th month report, Sakarya University, TURKEY http://www.sahakk.sakarya.edu.tr/hybridbus/index.html

[21] Koebl, M.; Elsener, M. & Kleeman, M. (2000). Urea-SCR: a promising technique to reduce NO*x* emissions from automotive diesel engines, *Catalysis Today* 59 (2000) 335-345.
[22] TRAD, (2010). http://www.trad.co.jp/english/products/egr1_1.html
[23] VALTRA, (2010). http://www.valtra.com/products/tractors/2346.asp

# 2

# Hourly Traffic Flow Predictions by Different ANN Models

Vedat TOPUZ
*The Marmara University*
*Istanbul,*
*Turkey*

## 1. Introduction

In this paper, different artificial neural network (ANN) models for predicting the hourly traffic flow in Second Tolled Bridge of Bosphorus in Turkey is performed. Second Bosphorus Bridge has an important role in urban traffic networks of Istanbul. The prediction of the traffic flow data is a vital requirement for advanced traffic management and traffic information systems, which aim to influence the traveler behaviors, reducing the traffic congestion, improving the mobility and enhancing the air quality. Detecting the trends and patterns in transportation data is a popular area of research in the business world to support the decision-making processes. The primary means of detecting trends and patterns has involved statistical methods such as clustering and regression analysis most of which use linear models. The goal is to predict a future value of data by using the past data set. On the other hand, non-linear approaches offer more powerful models, since they give more possibilities in the choice of the input-output relation.

ANNs approaches have been applied to a large number of problems because of their non-linear system modeling capability by learning ability using collected data. They offer highly parallel, adaptive models that can be trained by the experience. During the last decade, ANN has been applied widely to prediction of the traffic data.

We compared the generalization performance of the different ANN models such as Multi Layer Perceptron (MLP), Radial Basis Function Network (RBF), Elman Recurrent Neural Networks (ERNN) and Non-linear Auto Regressive and eXogenous input (NARX) type.

## 2. Traffic flow prediction

Detecting the trends and patterns in transportation data is a popular area of research in the business world to support the decision-making processes (Topuz, 2007). According to the theory, the traffic flow forecasting research can be divided into two types. One type is determined based on the mathematical model and second type is knowledge-based intelligent model of forecasting methods. Short-term traffic flow prediction is more influenced by the stochastic interferential factors than the long-term one, the uncertainty is greater and the disciplinarian laws are less obvious. Thus using the short-term traffic prediction models based on the classical mathematical methods, the precision of forecast cannot satisfactorily meet the demand of real-time traffic control systems (Su & YU, 2007).

Because the traffic system is a complex and variable system that involves a great deal of people, the traffic flow state has high randomness and uncertainty. The traditional traffic flow forecast methods such as Kalman filter, Moving Average (MA), Autoregressive Integrated Moving Average (ARIMA) and etc had been unable to satisfy the demand of the forecast precision that was increasing in practice (CHEN & MA, 2009)
On the other hand, artificial neural networks (ANNs) have been applied to a large number of problems because of their non-linear system modeling capacity by learning ability using collected data. During the last decade ANNs have been applied widely to prediction of the traffic data (SADEK, 2007). Different ANN approach such as MLP (Smith &. Demetsky, 1994), RBF (Celikoglu & Cigizoglu, 2006), ERNN (Li & Lu, 2009), Neuro-Genetic (Abdulhai et al. 2002) have been used to predict the traffic follow.
Such a prediction study including the NARX type ANN has been completed in this paper, to compare the effectiveness of different ANN approach.

## 3. Artificial Neural Network (ANN)

ANNs are designed to mimic the characteristics of the biological neurons in the human brain and nervous system. Given a sample vectors, ANNs is able to map the relationship between input and output; they "learn" this relationship, and store it into their parameters.. The training algorithm adjusts the connection weights (synapses) iteratively learning typically occurs through the training. When the network is adequately trained, it is able to generalize relevant output for a set of input data. They have been applied to a large number of problems because of their non-linear system modeling capacity.

### 3.1 Multi Layer Perceptron (MLP)

There are different types of ANN although the most commonly used architecture of ANN is the multilayer perceptron (MLP). MLP has been extensively used in many transportation applications due to its simplicity and ability to perform nonlinear pattern classification and function approximation. Therefore, it is considered the most commonly implemented network topology by many researchers (Transportation Research Board 2007).
MLP means a feedforward network with one or more layers of nodes between the input and output nodes and it is capable of approximating arbitrary functions. Two important characteristics of the MLP are: its nonlinear processing elements (activation function) which have a nonlinearity and their massive interconnectivity (weights).Typical MLP network is arranged in layers of neurons, where each neuron in a layer computes the sum of its inputs and passes this sum through an activation function (*f*). For this context designed MLP network is shown in figure 1.
The MLP network is trained with back-propagation, the first step is propagating the inputs towards the forward layers through the network. For a three-layer feed-forward network, training process is initiated from the input layer (Hagan 1996).

$$
\begin{aligned}
a^0 &= u \\
a^{m+1} &= f^{m+1}(W^{m+1}a^m + b^{m+1}), \quad m = 0,1 \\
y &= a^3
\end{aligned}
\tag{1}
$$

Where, $y$ is output vector, u is input vector, $f$ (.) is the activation function, W is weighting coefficients matrices, b is bias factor vector and m is the layer index. These matrices defined as;

$$W^1 = \begin{bmatrix} w_{1,1} & w_{1,2} & \dots & w_{1,S0} \\ w_{2,1} & w_{2,2} & \dots & . \\ . & . & \dots & w_{2,S0} \\ w_{S1,1} & w_{S1,2} & \dots & w_{S1,S0} \end{bmatrix}, W^2 = [w_{1,1}\ w_{1,2} \dots w_{1,S1}],\ b^1 = [b_1\ b_2\ ..\ b_{S1}]^T,\ b^2 = [b_1],$$

Here $S_0$ and $S_1$ are size of network input and hidden layer

In this study, sigmoid tangent activation functions are used in the hidden layer and linear activation function is used in the output layer respectively. These functions are defined as follows;

$$f^1 = \frac{(exp^n - exp^{-n})}{(exp^n + exp^{-n})},\ f^2 = n \tag{2}$$

The total network output is;

$$n_i^m = \sum_{j=1}^{s^{m-1}} w_{i,j}^m a_j^{m-1} + b_i^m \tag{3}$$

Second step is propagating the sensibilities (d) from the last layer to the first layer through the network: $d^3, d^2, d^1$. The error (e) calculated for output neurons is propagated to the backward through the weighting factors of the network. It can be expressed in matrix form as follows:

$$\begin{aligned} d^3 &= -2\dot{F}^3(n^3)(e) \\ d^m &= \dot{F}^m(n^m)(W^{m+1})^T d^{m+1},\ \text{for}\ \ m = 2, 1 \end{aligned} \tag{4}$$

Here $\dot{F}^m(n^m)$ is Jacobian matrix defined as follow;

$$\dot{F}^m(n^m) = \begin{bmatrix} \frac{\partial f^m(n_1^m)}{\partial n_1^m} & 0 & 0 \\ 0 & \frac{\partial f^m(n_2^m)}{\partial n_2^m} & 0 \\ 0 & 0 & \frac{\partial f^m(n_{s^3}^m)}{\partial n_{s^3}^m} \end{bmatrix} \tag{5}$$

*e* is mean square error,

$$e = \frac{1}{2}\sum_{\gamma=1}^{q}(y^\gamma - \hat{y}^\gamma)^2 \tag{6}$$

Where $\gamma$ is the sample in dimension $q$.

The last step in back-propagation is updating the weighting coefficients. The state of the network always changes in such a way that the output follows the error curve of the network towards down.

$$W^m(k+1) = W^m(k) - \alpha d^m (a^{m-1})^T$$
$$b^m(k+1) = b^m(k) - \alpha d^m \tag{7}$$

where α represents the training rate, k represents the epoch number. By the algorithmic approach known as gradient descent algorithm using approximate steepest descent rule, the error is decreased repeatedly

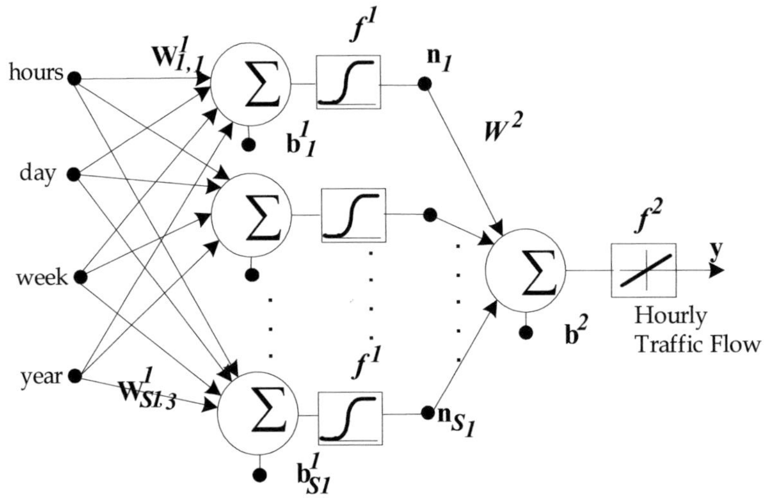


Fig. 1. Designed MLP Network Schematic Diagram.

### 3.2 Elman Recurrent Neural Networks (ERNN)

ERNN is also known partially recurrent neural network which is subclass of recurrent networks. It is MLP network augmented with additional context layers ($W^0$), storing output values (**y**), of one of the layers delayed ($z^{-1}$) by one-step and used for activating this other layer in the next time (t) step. The self-connections of the context nodes make it also sensitive to the history of input data, which is very useful in dynamic system modeling (Elman, 1990).

$$\mathbf{y}_{(t+1)} = f^1(\mathbf{W}^1\mathbf{x} + \mathbf{b}^1) + \mathbf{y}_{(t)}\mathbf{W}^0 \tag{8}$$

While ERNN use identical training algorithm as MLP, context layer weight ($\mathbf{W}^0$) is not updated as in equation 8. Schematic diagram of designed ERNN network is given in figure 2.

The ERNN network can be trained with any learning algorithm that is applicable to MLP such as backpropagation that is given above.

### 3.3 Radial Basis Function Network (RBF)

RBF network has a feed-forward structure consisting of two layers, nonlinear hidden layer and linear output which is given in figure in 3. RBF networks are being used for function approximation, pattern recognition, and time series prediction problems. Their simple structure enables learning in stages, and gives a reduction in the training time.

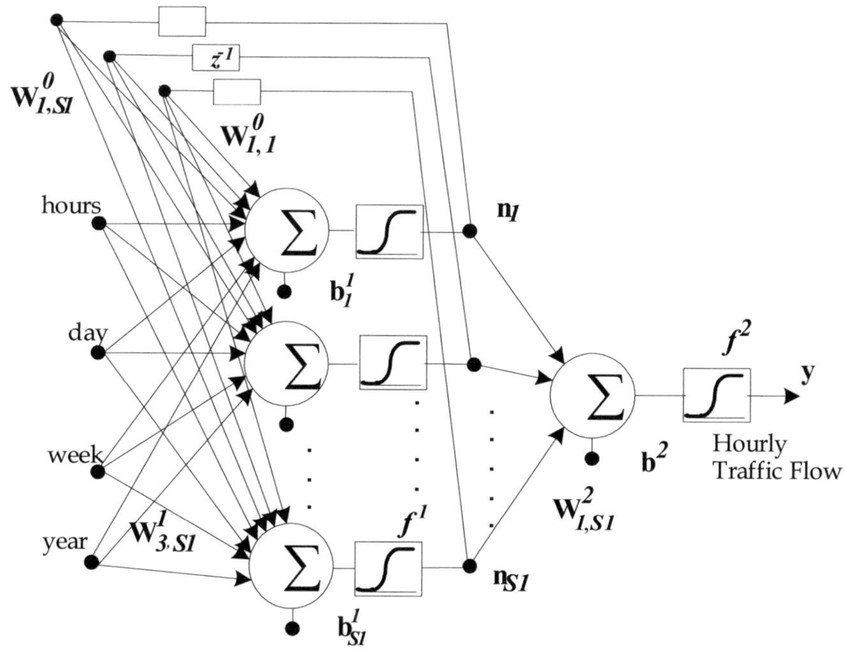


Fig. 2. Designed ERNN Network Schematic Diagram

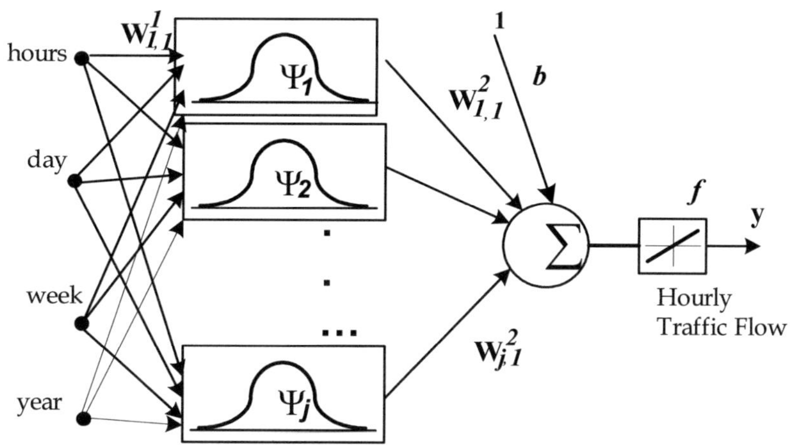


Fig. 3. Designed RBF Network Schematic Diagram.

The proposed model uses Gaussian kernel (Ψ) as the hidden layer activation function. The output layer implements a linear combiner of the basis function responses defined as; (Haykin, 1994)

$$y = b + \sum_{j=1}^{q} w_{i,j} \Psi_j \tag{9}$$

Where, $q$ is the sample size, $\Psi_j$ is response of the jth hidden neuron described as;

$$\Psi j = \exp\left[-\frac{\left\|\mathbf{x} - c_j\right\|^2}{2\sigma_j^{\ 2}}\right] \tag{10}$$

Where, $c_j$ is Gaussian function center value, and $\sigma_j$ is its variance.
RBF network training has two-stage procedure. In the first stage, the input data set is used to determine the center locations ($c_j$) using unsupervised clustering algorithm such as the *K*-means algorithm and choose the radii ($\sigma_j$) by the *k*-nearest neighbor rule. The second step is to update the weights (**W**) of the output layer, while keeping the ($c_j$) and ($\sigma_j$) are fixed.

### 3.4 Non-linear Auto Regressive and eXogenous Input type ANN (NARX)

A simple way to introduce dynamics into MLP network consists of using an input vector composed of past values of the system inputs and outputs. This the way by which the MLP can be interpreted as a NARX model of the system. This way of introducing dynamics into a static network has the advantage of being simple to implement. To deduce the dynamic model of realized system, NARX type ANN model can be represented as follows (Maria & Barreto, 2008);

$$y(k+1) = f_{ANN}(y(k), y(k-1)..., y(k-n+1), u(k), u(k-1)..., u(k-m+1)) + \varepsilon(k) \tag{11}$$

where $y(k{+}1)$ is model predicted output, $f_{ANN}$ is a non-linear function describing the system behavior, $u(k)$, $y(k)$, $\varepsilon(k)$ are input, output and approximation error vectors at the time instances $k$, $n$ and $m$ the orders of $y(k)$ and $u(k)$ respectively. Order of the process can be estimated from experience. Modeling by ANN relies on the consideration of an approximate function of $f_{ANN}$. Approximate dynamic model is constructed by adjusting a set of connection weight (**W**) and biases (**b**) via training function defined as MLP network. The NARX network, it can be carried out in one out of two modes:
***Series-Parallel (SP) Mode***: In this case, the output's regressor is formed only by actual values of the system's output defined as;

$$y(\hat{k}+1) = \hat{f}_{ANN}(y(k), y(k-1)..., y(k-n+1), u(k), u(k-1)..., u(k-m+1)) + \varepsilon(k) \tag{12}$$

Figure 4 shows the topology of one-hidden-layer NARX network when trained in the SP-mode.
***Parallel (P) Mode***: In this case, estimated outputs are fed back and included in the output's regressor defined as follows;

$$y(\hat{k}+1) = \hat{f}_{ANN}(y(\hat{k}), y(\hat{k}-1)..., y(k-\hat{n}+1), u(k), u(k-1)..., u(k-m+1)) + \varepsilon(k) \tag{13}$$

While NARX-SP type network is used in the training phase, NARX -P type network isused in the testing phase, which is given in figure 5.

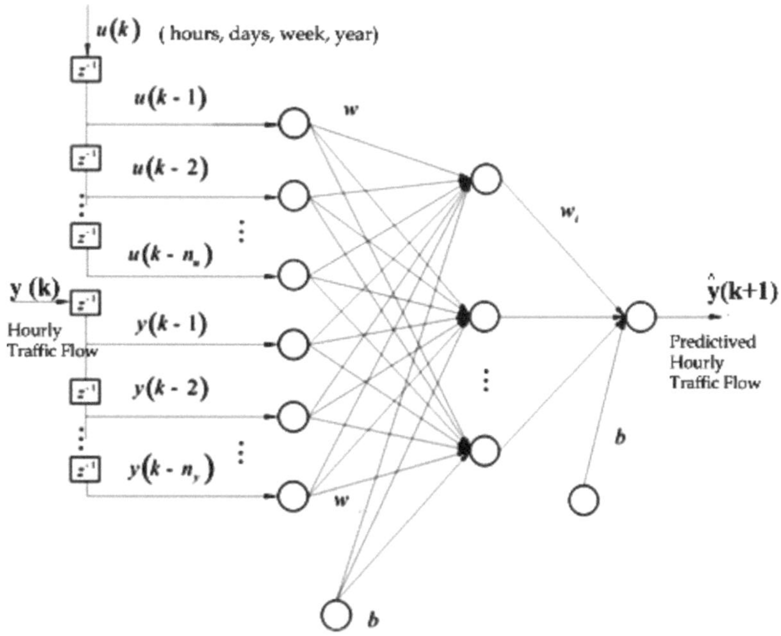


Fig. 4. Architecture of the NARX network during training in the SP-mode

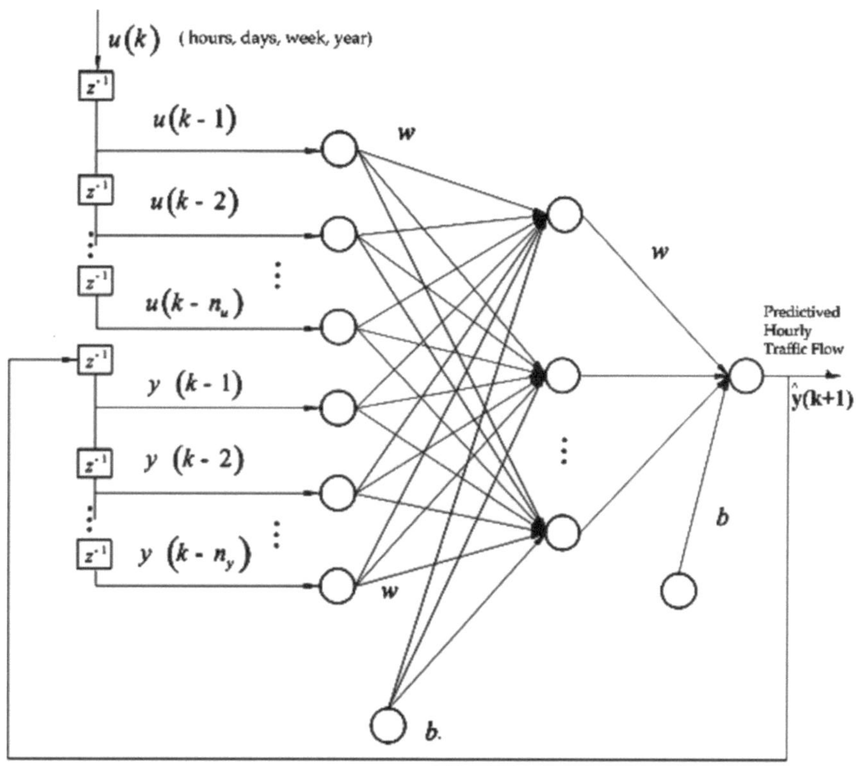


Fig. 5. Architecture of the NARX network during testing in the P-mode.

## 4. Designing process

ANN designing process involves four steps. These are gathering the traffic data, selecting the ANN architecture, training the network, and testing the network. In the training, four input variables have been used. These are, hours of the day: 0-23; days of the month: 1-30; days of the week: 1-7; and years: 2001-2003. Selecting the daily traffic demands as distinct variables both for days of the month and days of the week are mandatory situation, because of the weekly and monthly variance of the daily traffic demands. In the training phase, the target variables have been used as 2160 data related to the volumes, or hourly passing number of the vehicles (demand) of 4th months (April) of the years 2001-2003. In the testing phase, 720 data related to the hourly demand of 4th month (April) of the year 2004, have been used to compare the actual and neural output of the system performance.

***Gathering the Traffic Data:*** Data recorded by the tolls has been used for training, testing and analysis. Only 2880 number of traffic demand data related to the hours of the days of 4th months of the years 2001-2004 has been taken into consideration (Transportation Survey 2004).

***Selecting the Best Network Architecture:*** The number of hidden layer and neurons in the hidden layer(s) play very important roles for ANNs and choice of these numbers depends on the application. Influenced by theoretical works proved that single hidden layer is sufficient for ANNs to approximate any complex nonlinear function with any desired accuracy. In addition, determining the optimal number of hidden neurons is still a question to deal with. Although there is no theoretical basis for selecting these parameters, a few systematic approaches are also reported but the most common way of determining the number of hidden neurons is still trial and error approach. The number of input nodes that depend on the number of the time lags is also very important for the performance of the NARX type ANN model. The value of the time lags has to be determined by trial and error effort. Radius of the center point in nonlinear activation function is great effect the performance of the RBF network. Therefore, in addition to hidden layer neuron number, radiuses of the center point in neurons are also determined.

***Training and testing the network:*** Designed MLP, ERNN and NARX network are trained with Levenberg-Marquardt backpropagation learning algorithms, which described in section 3. MATLAB with Neural network toolbox is used for all ANN application (Demuth & Beale 2002).

The optimum learning, $\eta$ and the momentum, $\alpha$, rates are determined after trials observing the RMSE and R obtained at the end of the testing stage. When a smaller $\eta$ is selected, the changes to the synaptic weights (W) in the network had been smaller. The momentum rates $\alpha$ prevented the learning process from terminating in a local minimum of the error surface. It is seen that $\eta$ and $\alpha$ should be decreased if the number of the input and output layers are increased. On the other hand, the iteration number increases by decreasing $\eta$ and $\alpha$ value.

Once trained, the network can be used for predicting (test phase) the output for any input vector from the input space. This is called the "generalization property" of the network. To show this property of trained networks, the same experiment is done with testing data set that are not involved in training data set. It is also used the early stopping strategy (validation phase) to improve the generalization properties of designed network. In this strategy, the validation data set is used and the error on the validation set is monitored during the training process. The validation error normally decreases during the initial phase of training. However, when the network begins to overfit the data, the error on the validation set typically begins to rise. When the validation error increases for a specified number of iterations, which we choose 10, the training process is stopped. Root mean square

error (RMSE) values and correlations coefficients (R) are selected to compare the performance of designed network, which is defined as follows;
Features may be correlated, that is, a change in one feature $Y$ is associated with a similar change in another feature $\hat{Y}$. A correlation coefficient can be calculated as;

$$R = \frac{\sum_{i=1}^{K}(x_i - m_1)(y_i - m_2)}{\sqrt{\sum_{i=1}^{K}(x_i - m_1)^2 \sum_{i=1}^{K}(y_i - m_2)^2}} \tag{12}$$

Here $m_1$ is the mean value of all the values $x_i$ of feature $Y$, and $m_2$ is the mean of all $y_i$ values of the $\hat{Y}$ feature. The correlation coefficient assumes values in the interval [-1,1]. When R= -1 there is a strong negative correlation between $Y$ and $\hat{Y}$, when R=1 there is a strong positive correlation, and when R=0 there is no correlation at all.
Root mean squared error is also defined as follows;

$$RMSE = \sqrt{\frac{1}{q}\sum_{i=1}^{q}(yi_i - \hat{y}i)^2} \tag{13}$$

Where q is sample size, $y_i$, $\hat{y}_i$ are network output and predicted output respectively.

## 5. Experiments and results

In this section, firstly the performance of all MLP, ERNN, RBF and NARX type ANN model are investigated. Then scattering diagram and traffic flow graphics are drawn for the best ANNs models.

### 5.1 MLP network experiments

As a mentioned above, the performance of MLP network is highly dependent on the hidden layer neuron number. Therefore, performances of MLP network is tested using 5, 10,15,25,50 and 100 neurons in the hidden layer. End of the training and testing procedure s obtained correlations coefficients (R) and MSE values are given in table 1.

| Model | Training | | Test | |
|---|---|---|---|---|
| | RMSE | R | RMSE | R |
| MLP-5* | 0.0186 | 0.95 | 0.0190 | 0.944 |
| MLP-10* | 0.0093 | 0.975 | 0.0099 | 0.971 |
| MLP-15* | 0.0094 | 0.9750 | 0.0102 | 0.970 |
| MLP-25* | 0.0063 | 0.983 | 0.0077 | 0.978 |
| **MLP-50*** | **0.0032** | **0.991** | **0.0054** | **0.984** |
| MLP-100* | 0.0030 | 0.992 | 0.0070 | 0.98 |

*The numbers show the hidden neuron

Table 1. Different MLP Network Models RMSE and Correlation Coefficients (R) Values at the Training and Testing Phase.

It could be easily figure out in this table that, MLP-50 network shows the best generalization performance. It is also realized that to use more than 50 neurons in the hidden layer are not increase the generalization ability of designed networks. The best MLP network model scattering diagram is shown in figure 6.

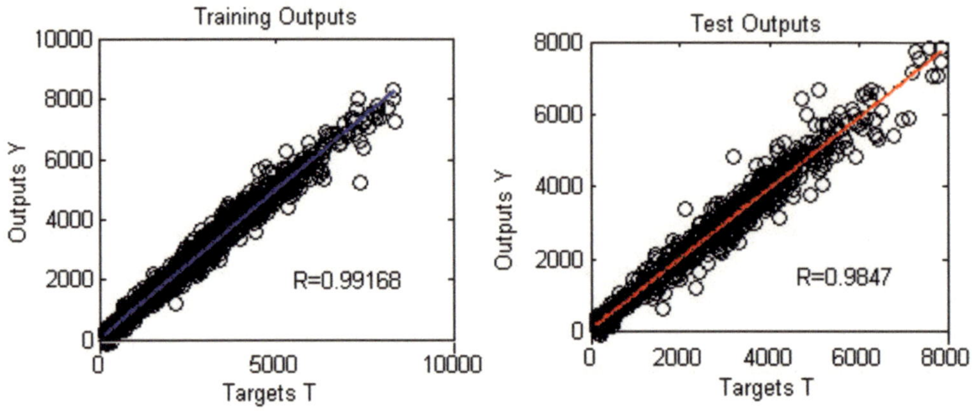


Fig. 6. MLP-50 Network Model Scattering Diagrams for Train and Test Output.

To show the obtained results, for only last week's predicted values with real traffic flow are given in figure 7. These figures with scattering diagrams show us that prediction errors are small enough to be accepted

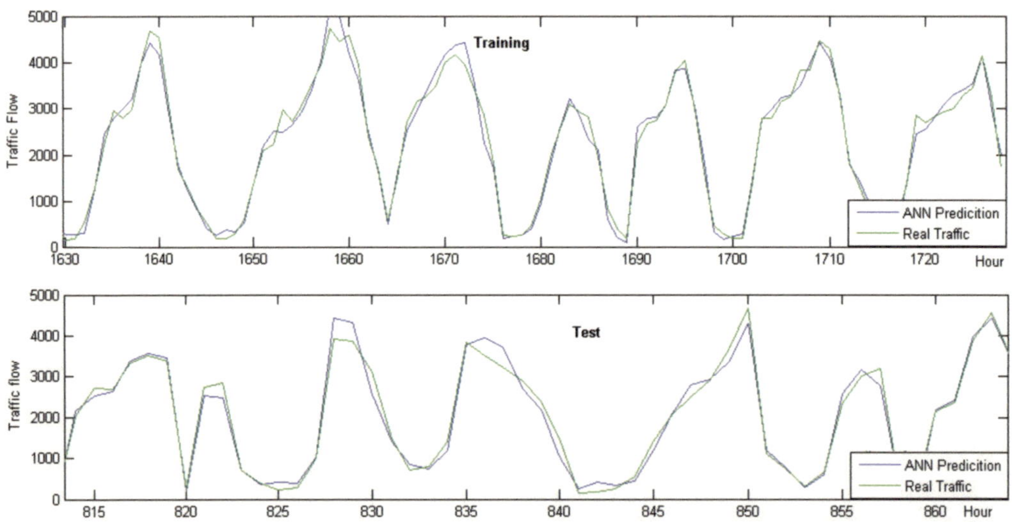


Fig. 7. The Best MLP Network Model Traffic Flow Graphics at the Train and Testing Phase.

### 5.2 ERN network experiments

The performance of ERNN is also dependent on the hidden layer neuron number. Hence, performances of ERNN is tested using 5, 10,15,25,50,100 neurons and obtained correlations coefficients (R) and MSE values are given in table 2. It could be easily figure out in this table that, ERNN -10 network shows the best generalization performance. The best ERNN model scattering diagram is shown in figure 8. Achieved the best ERNN results, with real traffic flow are given in figure 7.

| Model | Training | | Test | |
|---|---|---|---|---|
| | RMSE | R | RMSE | R |
| ERNN -5* | 0.0590 | 0.832 | 0.0527 | 0.839 |
| **ERNN -10** | **0.0556** | **0.844** | **0.0507** | **0.846** |
| ERNN -15 | 0.0754 | 0.778 | 0.0667 | 0.790 |
| ERNN -25 | 0.0924 | 0.730 | 0.0827 | 0.741 |
| ERNN -50 | 0.0872 | 0.740 | 0.0789 | 0.749 |
| ERNN -100 | 0.0936 | 0.715 | 0.0820 | 0.734 |

*The numbers show the hidden neuron.

Table 2. Different ERNN Models RMSE and Correlation Coefficients (R) Values at the Training and Testing Phase.

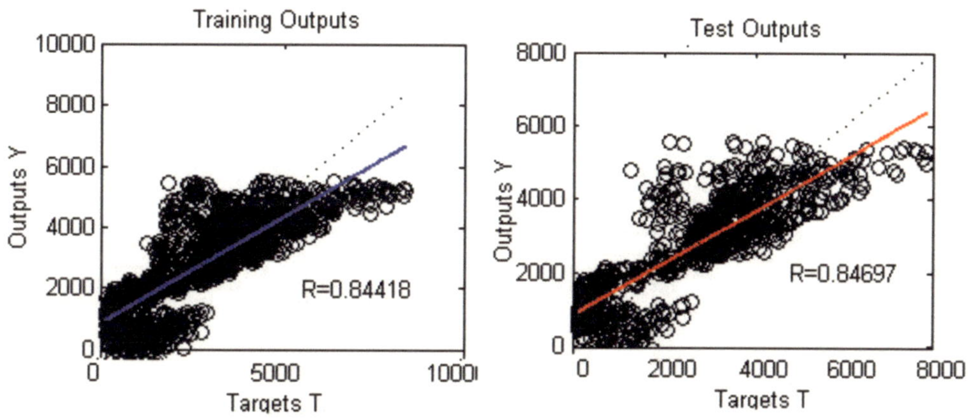


Fig. 8. ERNN -10 Network Scattering Diagrams for Train and Test Output.

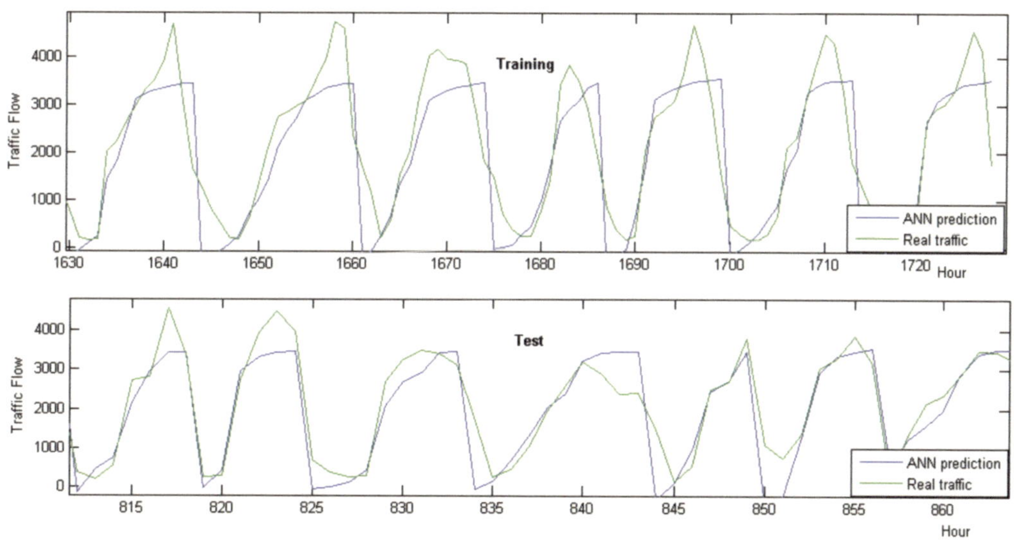


Fig. 9. The Best ERNN Model Traffic Flow Graphics at the Train and Testing Phase

### 5.3 RBF network experiments

Because of the performance of RBF network is depending on the radius of the center points and hidden layer neuron number, different RBF network models are tested. Obtained correlations coefficients (R) and MSE values are given in table 3. It could be easily figure out in this table that, RBF N250/S2 network shows the best generalization performance. The best RBF network model scattering diagram is shown in figure 10. Related traffic flow graphics on the test and training data is also given in figure 11.

| Model | Training | | Test | |
|---|---|---|---|---|
| | RMSE | R | RMSE | R |
| RBF N50/S2* | 0.0148 | 0.96 | 0.0155 | 0.957 |
| RBF N50/S3 | 0.0149 | 0.96 | 0.0172 | 0.953 |
| RBF N50/S3.5 | 0.0161 | 0.956 | 0.0183 | 0.949 |
| RBF N100/S2 | 0.0096 | 0.974 | 0.0123 | 0.966 |
| RBF N100/S3 | 0.0118 | 0.968 | 0.0152 | 0.958 |
| RBF N100/S3.5 | 0.0115 | 0.969 | 0.0148 | 0.959 |
| **RBF N250/S2** | **0.0066** | **0.982** | **0.0108** | **0.971** |
| RBF N250/S3 | 0.0096 | 0.974 | 0.0144 | 0.961 |
| RBF N250/S3.5 | 0.0102 | 0.973 | 0.0150 | 0.959 |
| RBF N1000/S2 | 0.0042 | 0.989 | 0.0145 | 0.961 |
| RBF N1000/S3 | 0.0074 | 0.980 | 0.0153 | 0.959 |

*N show the hidden neuron, S show the radius of the center points

Table 3. Different RBF Network Models RMSE and Correlation Coefficients (R) Values at the Training and Testing Phase.

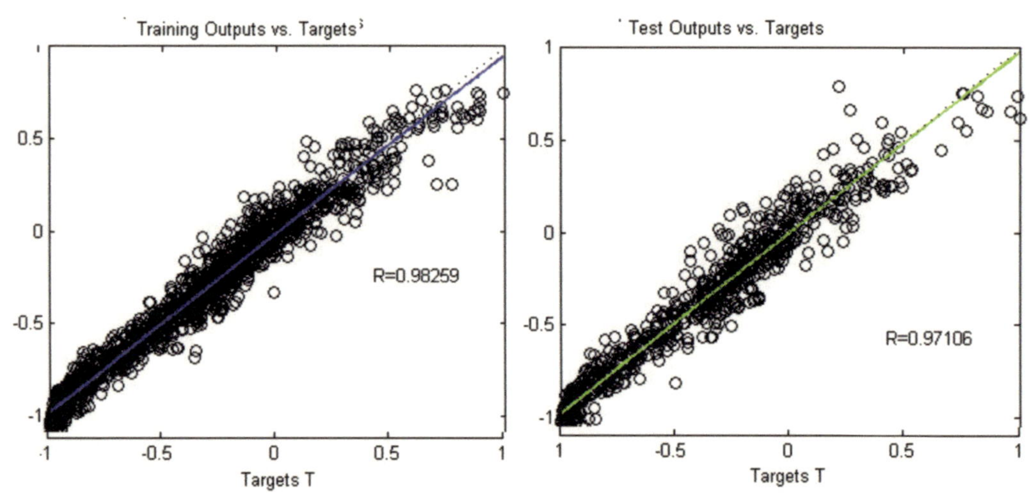


Fig. 10. RBF N250/S2 -10 Network Scattering Diagrams for Train and Test Output.

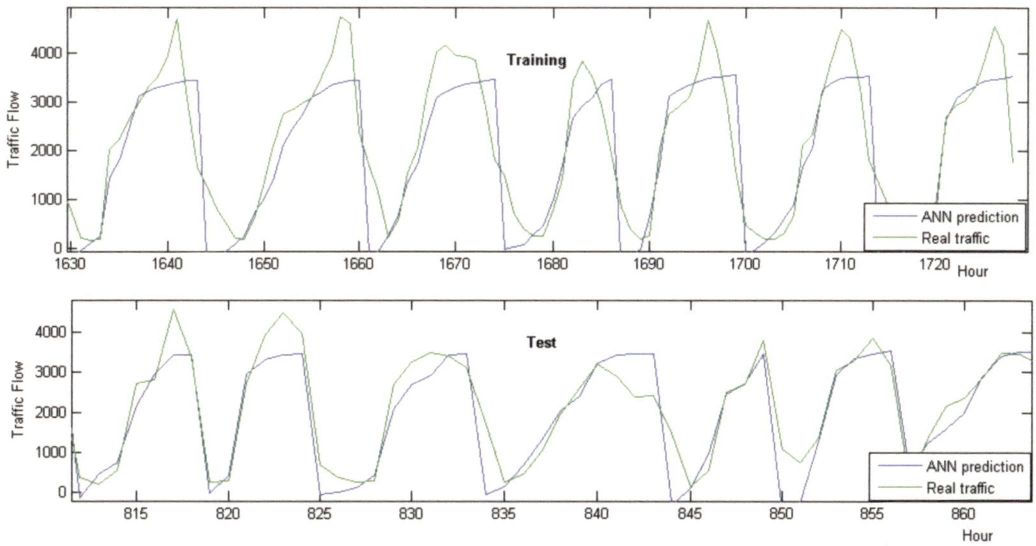


Fig. 11. Best RBF Network Model Traffic Flow Graphics at the Train and Testing Phase

### 5.4 NARX network experiments

The number of input nodes to the NARX type ANN model is dependent on the number of the time lags that is determined by the analysis of the auto-correlation function belonging to examined time series data set. Four time lags are found sufficient, hence; the input layer consisted of four nodes. Because the second step is largely a trial-and-error process and experiments involving neural networks with more than 4 input nodes and more than 100 neurons in the hidden layer didn't show any improvement in prediction accuracy. Obtained correlations coefficients (R) and MSE values are given in table 4, The best NARX network model scattering diagram is shown in figure 12.

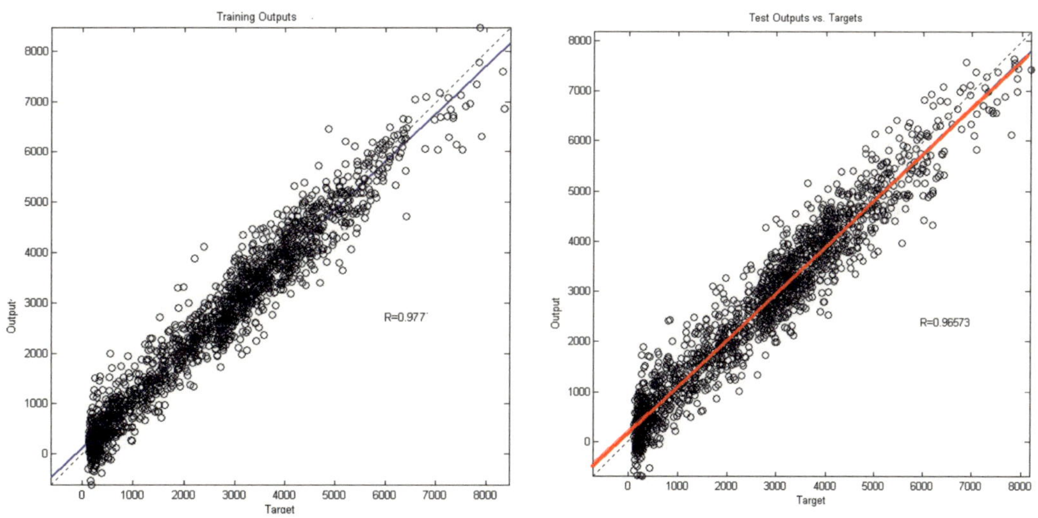


Fig. 12. NARX-H100D4 Network Scattering Diagrams for Train and Test.

| Model | Training | | Test | |
|---|---|---|---|---|
| | RMSE | R | RMSE | R |
| NARX-H5D5 | 0.0693 | 0.791 | 0.0750 | 0.749 |
| NARX-H10D5 | 0.0632 | 0.814 | 0.0627 | 0.839 |
| NARX-H15D5 | 0.0591 | 0.833 | 0.0554 | 0.849 |
| NARX-H25D5 | 0.0457 | 0.874 | 0.0527 | 0.867 |
| NARX-H50D5 | 0.0310 | 0.916 | 0.0327 | 0.911 |
| NARX-H100D5 | 0.0090 | 0.975 | 0.0109 | 0.960 |
| NARX-H5D10 | 0.0998 | 0.696 | 0.0872 | 0.713 |
| NARX-H10D10 | 0.0897 | 0.718 | 0.0820 | 0.734 |
| NARX-H15D10 | 0.0816 | 0.747 | 0.0789 | 0.749 |
| NARX-H25D10 | 0.0646 | 0.813 | 0.0527 | 0.839 |
| NARX-H50D10 | 0.0360 | 0.898 | 0.0507 | 0.846 |
| NARX-H100D10 | 0.0118 | 0.968 | 0.0212 | 0.958 |
| NARX-H100D20 | 0.0184 | 0.948 | 0.0190 | 0.944 |
| NARX-H100D2 | 0.0083 | 0.977 | 0.0153 | 0.959 |
| NARX-H100D3 | 0.0088 | 0.975 | 0.0144 | 0.961 |
| **NARX-H100D4** | **0.0083** | **0.977** | **0.0127** | **0.965** |

Table 4. Different NARX Network Models RMSE and Correlation Coefficients (R) Values at the Training and Testing Phase. (H hidden neuron number D delay time)

Achieved predicted values with real traffic flow are given in figure13. These figures with scattering diagrams show us that prediction errors are small enough to be accepted.

## 6. Conclusion

In this paper, different ANN models have predicted daily traffic demand in Second Tolled Bridge of Bosphorus, which has an important role in urban traffic networks of Istanbul. To find the best-predicted model feed forward type (MLP, RBF), recurrent type (ERNN) and NARX type ANNs are used. Previously every model parameters are investigated for each network type. In this stage, root mean square errors (RMSE) and correlation coefficient (R) values are used as a performance criterion.

Concerning neural networks that are created and simulated there have been excluded the below conclusions:

- The most efficient training algorithm proved to be the Levenberg-Marquardt Backpropagation, as it surpassed the others not only in quality of results but also in speed.
- For the MLP, ERNN and NARX type network, the best type of normalization proved to be the one in region -1 -1 in combination with sigmoid tangent activation function in the hidden layer nodes. For the RBF network, the best radius of the center points is found 2. For a NARX network, suitable choice of input delay is 4
- The number of nodes used in the hidden layer depends from the type and the complexity of the time series, and from the number of inputs.
- The results have shown that feed forward type network models have advantages over recurrent type network.
- It is also concluded that, many other transportation data prediction studies can be implemented easily and successfully by using the different ANN architectures.

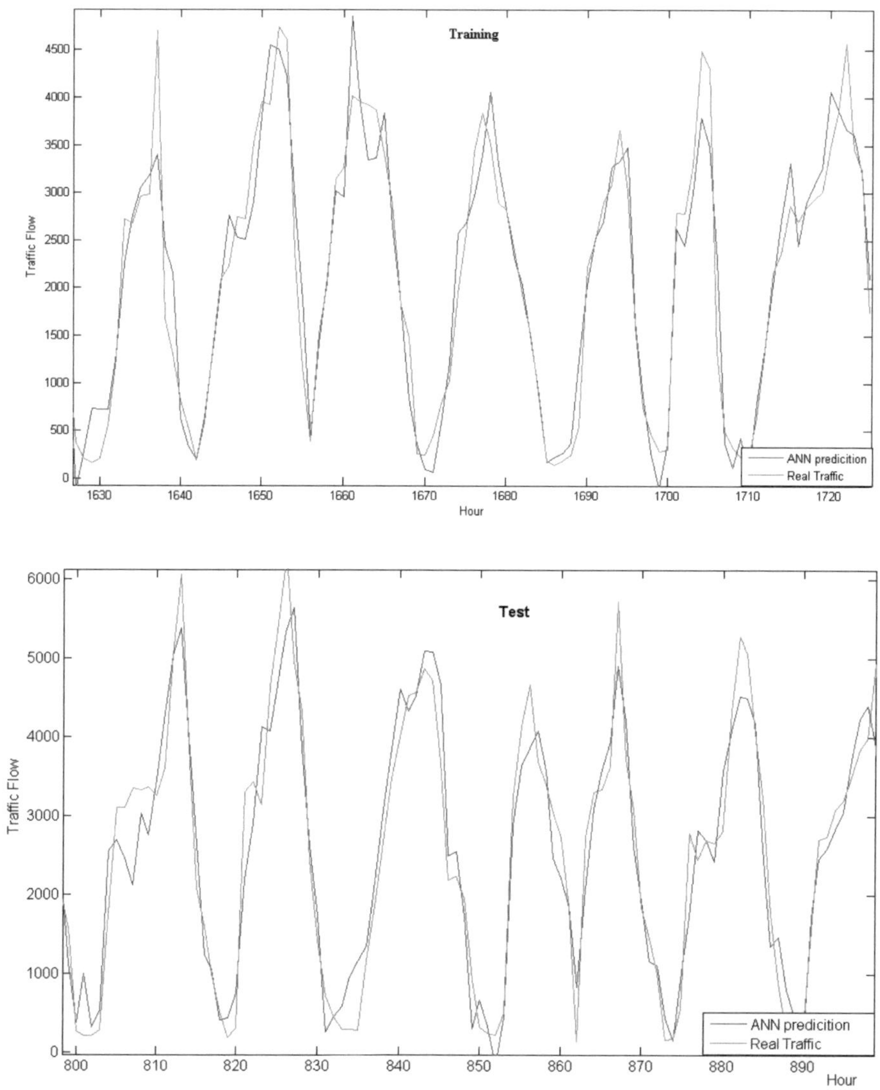


Fig. 13. NARX-H100D4 Network l Traffic Flow Graphics at the Train and Testing Phase

## 7. References

TOPUZ , V. (2007). Traffic Demand Prediction Using ANN Simulator, In: KES 2007/WIRN 2007, Part I, LNAI 4692, B. Apolloni et al. (Ed.), pp. (864-870), Springer-Verlag, Berlin Heidelberg

[Su, H & Yu, S. (2007). Hybrid GA Based Online Support Vector Machine Model for Short-Term Traffic Flow Forecasting In: APPT 2007, LNCS 4847, M. Xu et al. (Ed.), pp.(743-752), Springer-Verlag Berlin Heidelberg.

CHEN, B.P. & MA, Z. (2009), Short-term Traffic Flow Prediction Based on ANFIS, International Conference on Communication Software and Networks ICCSN. 2009, pp. 791-793, ISBN 978-0-7695-3522-7, IEEE.

SADEK, A, W. (2007), Artificial Intelligence Applications in Transportation, In: Transportation Research E-C113. Artificial Intelligence in Transportation Information for Application, pp.(1-6). Transportation Research Board, ISSN 0097-8515

Smith, B.L.& Demetsky, M.J. (1994): Short-term Traffic Flow Prediction: Neural Network Approach.Transp. Res. Rec. 1453, 98–104 .

Celikoglu, H.B & Cigizoglu, H.K (2007) Modelling public transport trips by radial basis function neuralNetworks In: Mathematical and Computer Modelling 45 (2007) 480–489 Elsevier.

Ruimin Li, R & Huapu Lu, H. Combined Neural Network Approach for Short-Term Urban Freeway Traffic Flow Prediction. Yu, H. He, and N. Zhang (Eds.): ISNN 2009, Part III, LNCS 5553, pp. 1017–1025, 2009. Springer-Verlag Berlin Heidelberg 2009

Abdulhai, B.& Himanshu,P & Will, R. (2002) 'Short-Term Traffic Flow Prediction Using Neuro-Genetic Algorithms', Journal of Intelligent Transportation Systems, 7: 1, 3 – 41

Transportation Research Board, Artifical Intelligence and Advanced Computing Applications Committee (January 2007) Artificial Intelligence in Transportation Information for Application Transportation Research Circular number E-C113 Washington, DC ISSN 0097-8515

Hagan, T. M. & Demuth, H.B & Beale, M., Neural Network Design, PWS Publishing Company, 1996, 2-44.

Elman, J. L. (1990) Finding Structure in Time; Cognitive Science, 14, 179-221.

Haykin, S. (1994) Neural Networks: A Comprehensive Foundation. Macmillan College Publishing Company, New York, USA. haykin p 300-305

Jose Maria P. Junior & Guilherme A. Barreto (2008) Narx reference: Long-Term Time Series Prediction with the NARX Network: An Empirical Evaluation In: Neurocomputing Volume 71, Issue 16-18 Pages: 3335-3343 ISSN:0925-2312

General Directorate Of Highways. Traffic and Transportation Survey 2000-2004. GDOH, Ankara, Turkey (2004)

Howard Demuth &Mark Beale (2002) Neural Network Toolbox User's Guide .The MathWorks, Inc. MA

# 3

# Local and Global Iterative Algorithms for Real-Time Short-term Traffic Flow Prediction

Eleni I. Vlahogianni, Ph.D and Matthew G. Karlaftis, Ph.D.
*National Technical University of Athens*
*Greece*

## 1. Introduction

When considering short-term prediction systems that operate in real-time and in an "intelligent" technology-based environment, the effectiveness depends, mostly, on predicting traffic information in a timely manner (Smith & Oswald, 2003). This implies that regardless of the traffic conditions met in real-time, a prediction system should not only be able to generate accurate single step ahead predictions of traffic flow, but also to operate on an iterative basis to produce reliable multiple steps ahead predictions in cases of data collection failures.

The bulk of research in short-term traffic flow prediction has concentrated on data-driven time-series models that construct the underlying rules of complex traffic datasets rather than working based on pre-determined mathematical rules; these models can be parametric - such as ARIMA models - or non-parametric (a review of the literature, methodologies and approaches used can be found in Vlahogianni et al., 2004). Among non-parametric approaches, neural networks and especially Multi-layer Feed-forward Perceptrons (MFLPs) have been shown to be most effective in forecasting traffic flow variables because of their propensity to account for a large range of traffic conditions and generate more accurate predictions than classical statistical forecasting algorithms (Smith & Oswald, 2003, Vlahogianni et al. 2004, Vlahogianni et al., 2005, Wang et al., 2006). Multi-layer Feed-forward Perceptrons (MFLPs) have memoryless neurons while input activation is a function only of the current input state and not of the past input-output relations and are very popular and widely used in time series traffic forecasting partly because of their ability to capture non-linear behavior in the data structures they model, regardless of their complexity (Hornik et al., 1989). They are usually trained under a pattern-based consideration; computations aim at distinguishing clusters that have different statistical properties (mean and variance).

However, the static operation of MLFP contradicts the temporal evolution of traffic flow. Short-term traffic flow prediction is significantly affected by the temporal evolution of traffic variables (Abdulhai et al., 2002, Stathopoulos & Karlaftis, 2003). The core research in traffic flow over the years has investigated a wide variety of nonlinear phenomena such as phase transitions (Kerner & Rehborn 1996), hysteretic effects (Hall et al. 1992, Zhang 1999), localized spatio-temporal behavior (Kerner 2002) and others. These phenomena - mainly observed in freeway bottlenecks - are difficult to predict and replicate in a simulation environment based on classical traffic flow theory (Kerner 2004). Moreover, recent efforts in

studying the temporal evolution of traffic flow in signalized arterials have indicated that traffic volume patterns - a sequence of volume states that define the time window for prediction – exhibit significant temporal variability and demonstrated that short-term traffic flow has variable deterministic and nonlinear characteristics that can be related to the prevailing traffic flow conditions (Vlahogianni et al. 2006a, 2008).
Although traffic flow theory has formulated improved theoretical bases in order to account for such phenomena - for example the three-phase traffic flow theory (Kerner 2004) - these effects are not taken into consideration in the process of short-term predictions using data-driven algorithms. Moreover, as indicated in most intelligent transportation system architectures, the dynamic nature of prediction algorithms are imperative to confront probable malfunctions in the data collection system or excesses in computational time (Smith & Oswald 2003, Vlahogianni et al. 2006b).
Iterative predictions provide the means to generate information on traffic's anticipated state with acceptable accuracy for a significant time horizon in cases of system failure. The present paper focuses on providing a comparative study between local and global iterative prediction techniques applied to traffic volume and occupancy time-series. The remainder of the paper is structured as follows: The concept of iterative predictions along with the basic notions of the proposed local and global prediction techniques in traffic flow prediction is presented in the following section. Next, the characteristics of the implementation area as well as the results of the comparative study are summarized. Finally, the paper ends with some concluding remarks.

## 2. Iterative predictions of traffic flow

Most predictions systems are dependent on data transmission. This suggests that continuous flow of volume and occupancy data is necessary to operate efficiently. However, it is common for most real-time traffic data collection systems to experience failures (Stathopoulos & Karlaftis 2003). For this, a real-time prediction system should be able to generate predictions for multiple steps ahead to ensure its operation in cases of data collection failures. The multiple steps ahead prediction problem can be formulated based on two conceptual approaches:

- *Direct approach*: Given a time-series of a variable - for example volume $V(t), V(t\text{-}1), \ldots$, a model is constructed to produce the state of a variable at $V(t+h)$ steps ahead, where $h > 1$.
- *Iterative or recursive approach*: Given a time-series of a variable a single step ahead model is constructed to produce a prediction $\hat{V}(t)$ at time $t$ that is then fed backwards to the network and is used as new input data in order to produce the next step $\hat{V}(t+1)$ prediction at $t$+1:

$$\hat{V}(t+1) = \left\{\hat{V}(t), V(t), V(t-1)\ldots\right\} \tag{1}$$

The direct approach to prediction has been implemented using neural networks in Vlahogianni et al. (2005). The reliability of direct prediction models is suspect because the model is forced to predict further ahead (Sauer 1993). This is the main argument in using iterative models in multiple steps ahead prediction. On the other hand, iterative predictions have the disadvantage of using the predicted value as input that is probably corrupted. Literature in short-term prediction of chaotic time-series indicates that the iterative approach to multiple steps ahead prediction is more accurate than direct one (Casdagli 1992).

Iterative approaches to neural network modeling have previously been applied to traffic flow predictions and results indicate that predictability decreases when the model uses previous predictions as inputs in the process of short-term predictions (Zhang 2000). This can be probably attributed to the fact that short-term traffic flow dynamics are not smooth and could not be captured by the simple structures of MLFP implemented so far (Vlahogianni et al. 2005). Two distinct categories of prediction techniques are further investigated; the local linear prediction techniques and the global neural networks. The next sub-sections provide the basics of the above techniques.

### 2.1 Local Weighted Linear model

The locally weighted linear (LWL) model finds similar local dynamics and produces a good estimate of the future state of a system, while neural networks are global approaches that attempt to build a single complex model for the entire range of behaviors identified in the time series. LWL belongs to the category of memory-based local prediction models. Locality in prediction is achieved by utilizing only the nearby states of the variable to be predicted. In brief, taken the case of predicting the next state of traffic volume $V(t+\tau)$ where $\tau$ is the embedding delay that matches the prediction step, the algorithm searches the $k$ nearest neighbors of $\mathbf{V}(t)$ i.e. the $k$ nearest states $\mathbf{V}(t')$ ($t'<t$) that minimize $\|\mathbf{V}(t) - \mathbf{V}(t')\|$, where $\|\cdot\|$ a metric (for example Euclidean) (Casdagli 1992). The local linear predictor results by fitting a linear polynomial to the pairs of $\mathbf{V}(t')$, $\mathbf{V}(t'+\tau)$; the contribution of each of the $k$ nearest neighbors is weighted by a weighting function (kernel) $K = e^{-\frac{d}{h}}$ -$d$ is the distance of a neighbor from the reference state and $h$ is the distance of the reference state and its furthest neighbor, among the $k$ nearest neighbors considered (Kononov 2007).

### 2.2 Temporal neural networks

The global iterative model to be used as an alternative theoretical approach to the proposed prediction framework is the temporal structures of neural networks. These networks are an extension of the static MLFPs that engage a memory mechanism to reconstruct the time-series under prediction in the $m$-dimensional Phase Space. The memory can be limited to the input layer or can extend to the entire neural networks - for example the hidden layer - and aims at creating a State-Space reconstruction process and converts the time series data of volume and occupancy in vectors; for example $\vec{V}_t = \left\{V_{t-\tau},\ldots,V_{t-(m-1)\tau}\right\}$, where $\tau$ is the time delay of and $m$ is the dimension the reconstructed volume in the State-Space. Networks that encompass tap delay memory structures only to the input layer of the form $\mathbf{y}(n) = [y(n), \ldots, y(n-m+1)]^T$ are known as time-delayed neural networks (TDNN), whereas networks that incorporate more complex memory mechanisms – for example a Gamma memory (de Vries & Principe 1992) – are known as time-lagged neural networks (TLNN).

The training of TLNN for iterative predictions feeds back the prediction at time $t$+1 and utilizes it as an input for the generation of next prediction step $t$ +2. For global models, this simple method is plagued by the accumulation of errors and the model can quickly diverge from the desired behavior (Principe et al. 2000). For this reason the training in the specific iterative neural network model is conducted via the temporal back-propagation algorithm known as Back-propagation to time (BPTT) (Webros 1990). BPTT are trajectory learning algorithms that create virtual networks by unfolding in time the original topologies on which back-propagation training is applied; the main objective is to improve the weights so as to minimize the total error over the whole time interval (Webros 1990). More specifically, all weights are duplicated spatially for an arbitrary number of time steps $\tau$; as such, each node that sends activation to the next has $\tau$ number of copies as well. For a training cycle $n$, the weight update is given by the following equation (Haykin 1999):

$$\mathbf{w}_{ji}(n+1) = \mathbf{w}_{ji}(n) + \eta\delta_j(n)\mathbf{x}_i(n) \tag{2}$$

where, $\mathbf{w}_{ji}(n+1)$ and $\mathbf{w}_{ji}(n)$ are the weights of the $i$-th synapse of the neuron $j$ at training cycle $n$+1 and $n$ respectively, $\eta$ is the learning rate, $\mathbf{x}_i(n)$ (i=1,2,...n) is the input vector and $\delta_j(n)$ is given by:

$$\delta_j(n) = \begin{cases} e_j(n)\phi'(\upsilon_j(n)),\ j \text{ neuron in the output layer} \\ \phi'(\upsilon_j(n))\sum_{r\in A} \mathbf{\Delta}_r^{\mathrm{T}}(n)\mathbf{w}_{rj},\ j \text{ neuron in the hidden layer} \end{cases} \tag{3}$$

where, $e_j(n)$ is the network's error, $\phi$ is the nonlinear activation function. Moreover, if $A$ is a set of all neurons whose inputs are fed by the $j$ neuron in the hidden layer is a forward manner, then $\upsilon_j(n) = \sum_{i=1}^{m} \mathbf{w}_{ji}\mathbf{x}_i(n) + b_j$ is the induced local field of neuron $r$ that belongs to the $A$ and $\mathbf{\Delta}_r^{\mathrm{T}}(n) = [\delta_r(n), \delta_r(n+1), ..., \delta_r(n+m)]^T$ is the local gradient vector.

Regardless of being static or dynamic, neural networks suffer from a usually manual trial-and-error process of optimization mainly with respect to their structure (number of hidden units) and learning parameters. On the other hand, automation in the optimization process of a neural network is most critical in complex neural structures. Recently, genetic algorithms have gained significant interest as they can be integrated to the neural network training to search for the optimal architecture without outside interference, thus eliminating the tedious trial and error work of manually finding an optimal network. Genetic algorithms are based on three distinct operations: selection, cross-over and mutation (Mitchell 1998); these operations run sequentially in order for a fitness criterion (in the specific case the minimization of the cross-validation error) to converge.

## 3. Implementation and results

The specific iterative prediction approach is implemented using 90-seconds volume and occupancy data collected along a signalized arterial in the centre of Athens (Greece) (Fig 1). Previous research on the area has demonstrated that traffic flow in the specific area - defined by the joint consideration of volume and occupancy – has a variable deterministic and nonlinear temporal evolution (Vlahogianni et al. 2008). The subsequent reconstruction of the series of volume and occupancy in the Phase-space using the mutual information criterion (Fraser & Swinney 1986) and the false nearest algorithm (Kennel et al. 1992) indicate that both volume and occupancy can be determined by a vector with time delay that equals to 1 and dimension that equals to 5.

Based on the above results, the recursive prediction techniques are set to have a look-back temporal window of 1-hour to retrieve and evaluate the traffic flow patterns. Moreover, for the recursive TDNN training, the neural network is set to be unfolded in an m=5 dimensional space in order to produce predictions. The depth of the Gamma of the TLNN was also calibrated to reflect the above reconstruction. Apart from the memory mechanism,

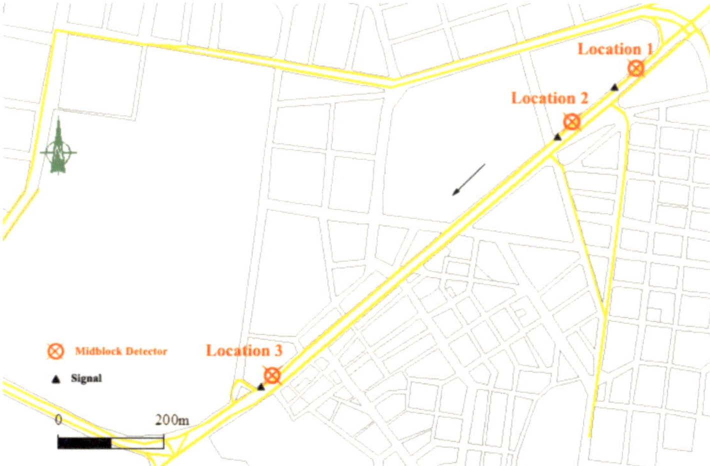


Fig. 1. Representation of the set of arterial links under study.

the difference between TDNN and TLNN implemented is that the second network extends the memory mechanism to the hidden layer too, in order to provide a fully non-stationary environment for the temporal processing of volume and occupancy series. The specifications regarding data separation as well as the genetic algorithm optimization are depicted on Table 1.

| Parameters | Specifications |
|---|---|
| Datasets: TR–CV–TE * | 60%-20%-20% |
| Levels | 1 hidden layer |
| Optimization | Genetic algorithm |
| Back-propagation | Genetic algorithm |
| Chromosome | $h \in [5,25]$ , $\gamma \in [0.01 - 0.1]$, $\mu \in [0.5 - 0.9]$ ** |
| Fitness function | Mean square error (cross-validation set) |
| Selection | Roulette |
| Cross-over | Two point ($p$=0.9) |
| Mutation | Probability $p$=0.09 |

** Training - Cross-validation - Testing*

*** h: neurons in hidden layer, $\gamma$: learning rate, $\mu$: momentum*

Table 1. Data and neural network specifications for iterative short-term volume and occupancy prediction.

The results of the comparative study are summarized in Table 2. As can be observed the TLNN performs significantly better - with regards to the mean relative percent error of prediction - than the local weighted linear model under the iterative prediction framework in both volume and occupancy. When compared to the iterative predictions of a TDNN, it is observed that TDNN performs comparable to the TLNN. However, as the same does not apply to the case of occupancy; further statistical investigation is conducted to the series of volume and occupancy in order to explain the behavior of the models regarding occupancy predictions. Results from a simple LM ARCH (Eagle 1982) that tests the null hypothesis of no ARCH effect lying in the data series of volume and occupancy shows that occupancy exhibits higher time-varying volatility than volume that is difficult to be captured.

| Iterative Models | Mean Relative Percent Error (%) | |
|---|---|---|
| | Volume | Occupancy |
| LWL | 21 | 30 |
| TDNN | 14 | 26 |
| TLNN | 13 | 22 |

Table 2. Prediction Results (Mean Relative Percent Error) of the comparative study.

Fig 2 and Fig 3 depict the relationship of the actual and the predicted values of volume and occupancy equally. A systematic error is observed in the predictions of volume using the local prediction model. Moreover, there seems to be a difficulty in predicting high volume values as observed in Fig. 2. As for the occupancy predictions, there seem to be much more scattered that the ones of volume; R2 values are lower than the ones of volume.

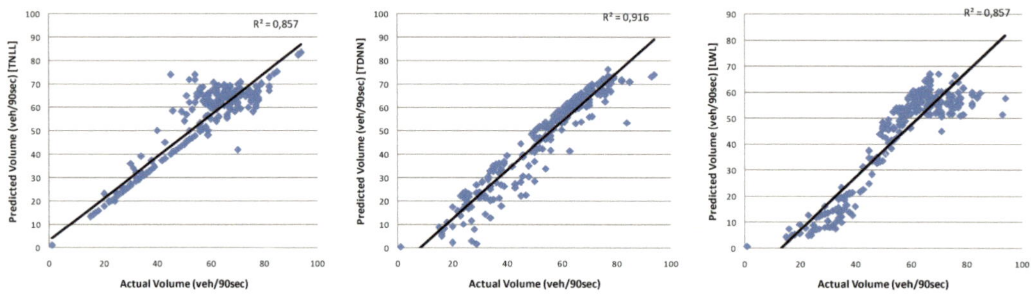


Fig. 2. Actual versus predicted values of traffic volume for the three iterative prediction techniques evaluated.

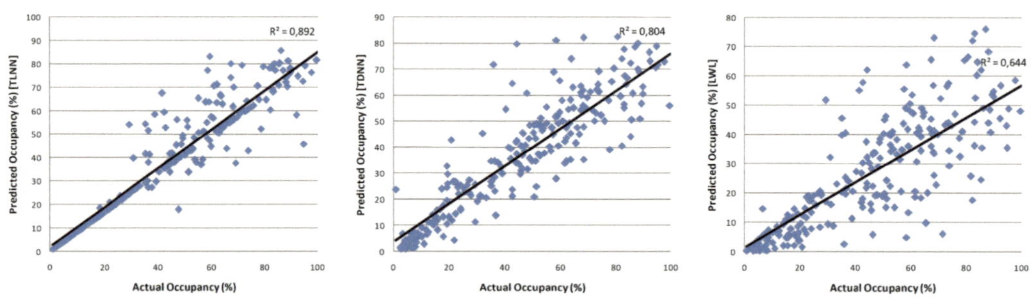


Fig. 3. Actual versus predicted values of occupancy for the three iterative prediction techniques evaluated.

In order to investigate the performance of the iterative models during the formation of congested conditions, two distinct time periods are selected for further studying the time series of the actual and predicted volume and occupancy with regards to different methodologies. These two periods depict the onset of the morning (Figure 4) and the afternoon peak (Fig. 5).

As can be observed, although iterative TLNN exhibited improved mean relative accuracy when compared to the iterative TDNN, both models seem to capture the temporal evolution of the two traffic variables under study. In the case of afternoon peak where the series of volume exhibit a oscillating behavior - in contrast to the trend observed in volume and occupancy during the onset of the morning peak, both neural network models either overestimate of under-estimate the anticipated values of traffic volume. As for the LWL model, predictions as depicted in the time series of the actual versus the predictive values of traffic volume and occupancy can be considered as unsuccessful.

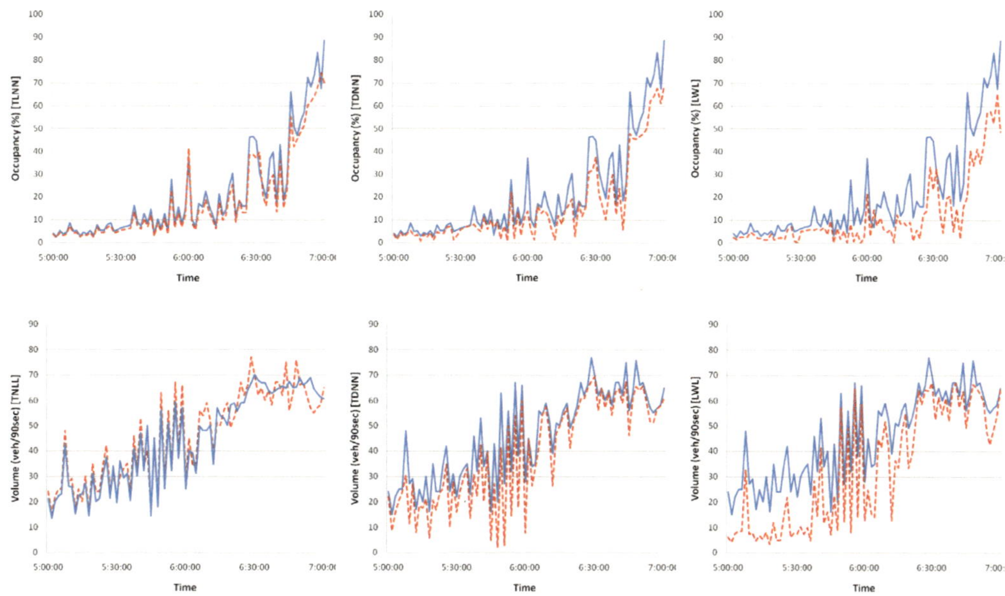


Fig. 4. Time-series of actual and predicted (dashed line) values of traffic volume (vh/90sec) for the onset of the morning peak.

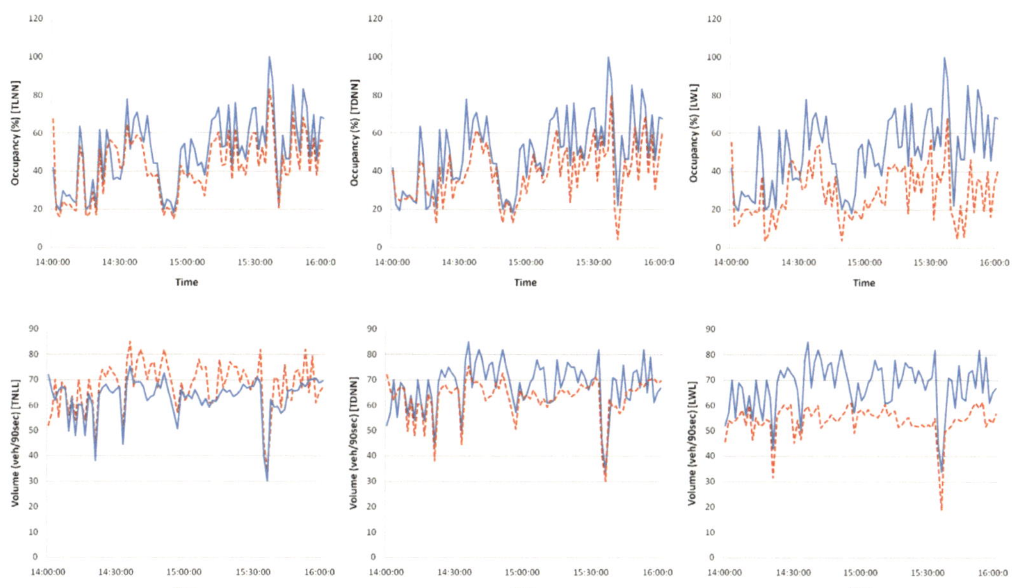


Fig. 5. Time-series of actual and predicted (dashed line) values of traffic volume (vh/90sec) for the onset of the afternoon peak.

## 4. Conclusions

Modern intelligent transportation systems require prediction algorithms that are adaptable and self-optimized in terms of the prevailing traffic flow conditions. Neural networks have been for long considered a prominent approach short-term prediction of traffic variables. The present paper extends past research by focusing on purely temporal structures of neural networks that provide iteratively short-term traffic flow predictions. A comparative study is conducted between local prediction techniques and neural networks with respect to the predictive accuracy. Results indicate that the global neural networks techniques outperform the local predictors, both when considering the mean behavior of the models and their behavior in critical traffic flow conditions, such as the onset of the morning and afternoon peak in signalized arterials. The optimal accuracy is attained by the TLNN that is the most complex temporal neural network among those tested.

From a conceptual standpoint, the TLNN implemented is fully compatible with the complex non-stationary features of traffic flow. From a methodological standpoint a central consideration should be kept in mind; as the aim is mainly at the real-time implementation, the extensive computational time to train and optimize such networks should be considered. It is evident that a retraining strategy is needed in order for the neural structures to incorporate and learn newly observed traffic flow events. Although the last is not required during the entire real-time operation of the model, research should be focus on the manner the accuracy of iterative predictions decreases over time, as well as the formulation of a mathematical or empirical criterion to evaluate the time neural networks should be retrained.

## 5. References


B. Abdulhai, H. Porwal and W. Recker, "Short-term traffic flow prediction using neuro-genetic algorithms," Intelligent Transportation Systems Journal, vol.7, no. 1, pp. 3–41, Jan. 2002.

M. Casdagli, "Chaos and deterministic versus stochastic non-linear modeling", Journal of the Royal Statistical Society. Series B (Methodological), Vol. 54, No. 2, pp. 303-328, 1992.

R.F. Engle, Autoregressive conditional heteroskedasticity with estimates of the variance of UK inflation. Econometrica 50, 987– 1008, 1982.

J. D. Farmer and J. J. Sidorowich, "Predicting Chaotic Time-Series", Phys. Rev. Lett., vol. 59, issue 8, pp. 845 - 848, Aug 1987.

A. M., Fraser and H. L., Swinney, (1986). Independent coordinates for strange attractors from mutual information, Physical Review A, 33 (2), pp. 1134-1140.

F. L Hall, V. F. Hurdle and J. H. Banks, "Synthesis of Occupancy (Or Density) Relationships on Freeways," Transportation Research Record 1365, TRB, National Research Council, Washington, DC, pp. 12-18, 1992.

S. Haykin, Neural Networks: A comprehensive foundation, Prentice Hall Upper Saddle River, NJ, 1999.

K. Hornik, M. Stinchcombe and H. White, "Multilayer feed-forward networks are universal approximators," Neural Networks, vol.2, no. 5, pp. 359–366, Jul. 1989.

M. B., Kennel, R., Brown and H. D. I., Abarbanel, Determining embedding dimension for phase-space reconstruction using a geometrical construction, Physics Reviews A, 45, 3403, 1992.

B. S. Kerner and H. Rehborn, "Experimental properties of complexity in traffic flow," Phys. Rev. E vol.53, no. 5, pp. R4275 - R4278 May 1996.

B. S. Kerner, "Empirical macroscopic features of spatial-temporal traffic patterns at highway bottlenecks," Physical Review E, vol.65, no. 4, pp. 046138 1-30, Apr. 2002.

B. S. Kerner, The Physics Of Traffic: Empirical Freeway Pattern Features, Engineering Applications, And Theory. Springer Verlag, New York., 2004.

E. Kononov, Visual Recurrence Analysis version 5, May 2007.

M. Mitchell. An introduction to genetic algorithms. MIT Press, ISBN: 0262631857, 1998.

N. Y. Nikolaev and H. Iba, Adaptive Learning of Polynomial Networks: Genetic Programming, Backpropagation and Bayesian Methods, Springer, NY, 2006.

J. C. Principe, N. R. Euliano and C. W. Lefebvre, Neural and Adaptive Systems: Fondamentals Through Simulations. John Wiley and Sons, Inc, 2000.

T. Sauer, Time series prediction by using delay coordinate embedding, in: Weigend A.S., Gershenfeld N.A. (Eds.), Time series prediction: Forecasting the future and understanding the past, Addison Wesley, Reading MA, 1993.

B. L. Smith and R. K. Oswald, "Meeting Real-Time Requirements with Imprecise Computations: A Case Study in Traffic Flow Forecasting," Computer Aided Civil and Infrastructure Engineering. vol. 18, no. 3, pp. 201-213, May 2003.

A. Stathopoulos and M. G., Karlaftis, "A multivariate state-space approach for urban traffic flow modelling and prediction," Transportation Research Part C, 11(2), 121-135, April 2003.

E. I. Vlahogianni, J. C. Golias and M. G. Karlaftis, "Short-Term Traffic Forecasting: Overview of Objectives and Methods," Transport Reviews, vol. 24, no. 5, pp. 533-557, Sept. 2004.

E. I. Vlahogianni, M. G. Karlaftis and J. C. Golias, "Optimized and meta-optimized neural networks for short-term traffic flow prediction: A genetic approach," Transportation Research C, vol.13, no. 3, pp. 211-234, June 2005.

E. I. Vlahogianni, M. G. Karlaftis and J. C. Golias, "Statistical Methods for Detecting Nonlinearity and Non-Stationarity in Univariate Short-term Time-series of Traffic Volume," Transportation Research Part C,14(5), 351-367, 2006.

E. I. Vlahogianni, M. G. Karlaftis, J. C. Golias, and N. D. Kourbelis, Pattern-based Short-term Urban Traffic Predictor. Proceedings of the 9th International IEEE Conference on Intelligent Transportation Systems, September, Torondo, Canada, 2006.

E. I. Vlahogianni, M. G. Karlaftis and J. C. Golias, Statistical Identification of Temporal Patterns in Short-term Urban Traffic Flow, Computer-Aided Civil and Infrastructure Engineering, accepted for publication, 2008.

B. de Vries and J. C. Principe, The gamma model—A new neural-net model for temporal processing, Neural Networks, 5, 565-576, 1992.

Y. Wang, M. Papageorgiou, and A. Messmer, "A Real-Time Freeway Network Traffic Surveillance Tool," IEEE Transactions on Control Systems Technology, vol.14, no. 1, pp. 18-31, Jan. 2006.

P. J.Webros, "Backpropagation Through time: What it does and How to do it", IEEE Proceedings Vol 78, No. 10, 1550-1567 Oct. 1990.
H. M. Zhang, "A mathematical theory of traffic hysteresis," Transportation Research Part B, vol.33, no. 1, pp. 1-23. Feb. 1999.
H. M. Zhang, "Recursive Prediction of Traffic Conditions With Neural Networks," Journal of Transportation Engineering, 126(6), 472–481, 2000.

# 4

# Computer Vision Techniques for Background Modelling in Urban Traffic Monitoring

José Manuel Milla, Sergio L. Toral, Manuel Vargas and Federico Barrero
*University of Seville*
*Spain*

## 1. Introduction

Traffic data collection is an essential issue for road-traffic control departments, which need real-time information for traffic-parameter estimation: road-traffic intensity, lane occupancy, congestion level, estimation of journey times, etc., as well as for early incident detection. This information can be used to improve road safety as well as to make an optimal use of the existing infrastructure or to estimate new infrastructure needs.

In an intelligent transportation system, traffic data may come from different kinds of sensors. The use of video cameras (many of which are already installed to survey road networks), coupled with computer vision techniques, offers an attractive alternative to other traffic sensors (Michalopoulos, 1991). For instance, they can provide powerful processing capabilities for vehicle tracking and classification, providing a non-invasive and easier to install alternative to traditional loop detectors (Fathy & Siyal, 1998; Ha et al., 2004).

Successful video-based systems for urban traffic monitoring must be adaptive to different traffic or environmental conditions (Zhu & Xu, 2000; Zhou et al., 2007). Key aspects to be considered are motion-based foreground/background segmentation (Piccardi, 2004; Beymer et al., 2007; Kanhere & Birchfield, 2008), shadow removal algorithms (Prati et al., 2003; Cucchiara et al., 2003), and mechanisms for providing relative robustness against progressive or sudden illumination changes. These video-based systems have to deal with specific difficulties in urban traffic environments, where dense traffic flow, stop-and-go motion profiles, vehicle queues at traffic lights or intersections, etc., would be expected to occur.

This chapter is focused on background subtraction, which is a very common technique for detecting moving objects from image sequences using a static camera. The idea consists of extracting moving objects as the foreground elements obtained from the "difference" image between each frame and the so-called background model of the scene (Spagnolo et al., 2006). This model is used as a reference image to be compared with each recorded image. Consequently, the background model must be an accurate representation of the scene after removing all the non-stationary elements. It must be permanently updated to take into account the eventual changes in the lighting conditions or in the own background contents. Surveys and comparisons of different algorithms for background subtraction can be found in the literature (Piccardi, 2004; Chalidabhongse, 2003; Cheung & Kamath, 2004).

Regarding to the category of parametric background subtraction algorithms, in the simplest case, it is assumed that each background pixel can be modelled by a single unimodal probability density function. This is the case of the algorithm known as *running Gaussian average* (Wren et al., 1997; Koller et al., 1994), which is a recursive algorithm where a Gaussian density function is fitted for each pixel.

Temporal median filter is another common strategy which has been reported to perform better than those methods based on the average. The background estimate is defined for each pixel as the median of all the recent values (in the case of the non-recursive version of the algorithm). The assumption is that a background pixel must be clearly visible for more than 50% of the considered period (Cucchiara et al., 2003; Lo & Velastin, 2001; Zhou & Aggarwal, 2001).

*Mixture of Gaussians (MoG)* is another parametric strategy that has also been widely used (Stauffer & Grimson, 1999; Stauffer & Grimson, 2000; Harville, 2002). A single Gaussian density function for each pixel is not enough to cope with non-stationary background objects, such as waving trees or other natural elements. The idea under the MoG is to be able to model several background objects for each pixel. The achieved background tries to model the different intensities that can appear on each background pixel, using a mixture of *n* Gaussian density functions (Power & Schoonees, 2002). The optimal tuning of the parameter set in this algorithm is considered not to be a trivial issue. In White & Shah (2007), an automatic tuning strategy based on particle swarm optimization is proposed.

Another set of algorithms lay in the category of non-parametric algorithms. They are more suitable when it is assumed that the density function is more complex or cannot be modelled parametrically, since a non-parametric approach is able to handle arbitrary density functions. *Kernel density estimation* (*KDE*) is an example of non-parametric methods. It tries to solve a problem with the MoG and the other previous methods. These previous methods are able to effectively describe scenes with smooth behaviour and limited variation, as in the case of gradually evolving scenes. However, in the presence of a dynamic scene with fast variations or non-stationary properties, the background cannot be accurately modeled with a set of Gaussians. This technique overcomes the problem by estimating background probabilities at each pixel from many recent samples using kernel density estimation (Elgammal et al., 1999). In Mittal & Paragios (2004), density functions are estimated in a higher-dimensional space combining intensity information with optical flow, in order to build a method able to detect objects that differ from the background in either motion or intensity properties.

Another non-parametric approach is followed by the algorithm based in the called *Codebook model* (Kim et al., 2005). In this case, the background model for each pixel is represented by a number of *codewords* (instead of parameters representing probabilistic functions) which are dynamically handled following a quantization/clustering technique. An important parallel issue in the conception of this technique is an appropriate colour modelling. Haritaoglu et al. (2000) describe what they call *W4* algorithm, where each background pixel is represented by a combination of the minimum and maximum values together with the maximum allowed change in two consecutive frames.

A different category of methods considers predictive strategies for modelling and predicting the state dynamics at each pixel. Some of them are based on Kalman filter (Karmann & Brandt, 1990; Koller et al., 1994), where intensity values and spatial derivatives are combined to form a single state space for background tracking. Alternatively, they may rely on the Wiener filter, as the *Wallflower algorithm* (Toyama et al., 1999), or on more complicate models such as autoregressive models (Monnet et al., 2003; Zhong & Sclaroff, 2003). Finally, we can also mention methods based on eigenspace representation, known as

eigenbackgrounds (Oliver et al., 2000), where new objects are detected by comparing the input image with an image reconstructed via the eigenspace.
Apart from background subtraction techniques, another extended approach is based on salient feature detection, clustering and tracking (Beymer & Malik, 1996; Coifman et al., 1998). In this case, no background model has to be estimated and continuously updated. Instead, a bunch of prominent features that are expected to be stable along time are extracted from the vehicles' image. Then, sophisticated spatiotemporal clustering algorithms are applied in order to group those features which are likely to belong to the same vehicle (proximity, motion coherence, velocity, can be used as clues). The main problem with these algorithms is that they assume that all the features for a given vehicle lie on the same plane, which can be acceptable for far viewpoints and small targets. Some other approaches try to overcome this problem projecting the extracted features onto a plane parallel to the road surface (Kanhere & Birchfield, 2008).
From an implementational point of view, video-based traffic equipments are frequently based on embedded processors with significant computational limitations. They have to perform several tasks in real time, including considerable amount of image processing (Toral et al., 2009a). In this chapter, background subtraction algorithms with low computational requirements are considered for implementation on embedded processors. In particular, algorithms that allow reducing floating point computations to a minimum are preferable. This is the case of the above-mentioned median filter. However, the computation of the median value for each pixel from a number of recent samples is also a costly operation. A recursive algorithm, based on the sigma-delta filter, providing a very fast and simple approximation of the median filter with the additional benefit of having very low memory requirements, was proposed by McFarlane & Schofield (1995). In this algorithm, the running estimate of the median is incremented by 1 if the input pixel is above the estimate and decreased by one if over it. Manzanera and Richefeu (2004) use a similar filter to compute the time-variance of the pixels, which is used for classifying pixels as "moving" or "stationary". Recent enhancements of this algorithm have been proposed by Manzanera and Richefeu (2007), with the addition of some interesting spatiotemporal processing, at the expense of a higher complexity.
In addition to the concern on computational efficiency, this chapter is specifically focused in urban traffic environments, where very challenging conditions for a background subtraction algorithm are common: dense traffic flow, eventual traffic congestions or vehicle queues are likely to appear. In this context, background subtraction algorithms must handle the moving objects that merged into the background due to a temporary stop and then become foreground again. Many background subtraction algorithms rely on a subsequent post-processing or foreground validation step, using object localization and tracking, in order to refine the foreground detection mask. The aim of the proposed algorithm is to avoid the need of this subsequent step, preventing the background model to incorporate these objects which are stopped for a time gap and maintaining them as part of the foreground. At the same time, the algorithm should avoid the background model to get too obsolete after a change in the true background or in the illumination conditions. Consequently, special attention must be paid in deciding when and how updating the background model, avoiding "pollution" of the model from foreground slow moving or stopped vehicles, while preventing, at the same time, the background model to get outdated.
A new background subtraction algorithm based on the sigma-delta filter is described in this chapter and then compared with previous versions reported in the literature. A more

reliable background model is achieved in common adverse conditions typical of urban traffic scenes, satisfying the goal of low computational requirements. Moreover, the implementation of the proposed algorithm on a prototype embedded system, based on an off-the-shelf multimedia processor, is discussed in this chapter. This prototype is used as a test-bench for comparison of the different background subtraction algorithms, in terms of segmentation quality performance and computational efficiency.

## 2. Sigma-Delta background estimation algorithms

### 2.1 Basic Sigma-Delta algorithm

The basic sigma-delta background estimation algorithm provides a recursive computation of a valid background model of the scene assuming that, at the pixel level, the background intensities are present most of the time. However, this model degrades quickly under slow or congested traffic conditions, due to the integration in the background model of pixel intensities belonging to the foreground vehicles. Table 1 describes the basic sigma-delta algorithm from Manzanera & Richefeu (2004) (a statistical justification of this method is given in Manzanera, 2007). For readability purposes, the syntax has been compacted in the sense that any operation involving an image should be interpreted as an operation for each individual pixel in that image.

```
M_0 = I_0                               // Initialize background model M
V_0 = 0                                 // Initialize variance V
for each frame t
    Δ_t = |M_t − I_t|                   // Compute current difference
    if Δ_t ≠ 0
        V_t = V_{t−1} + sgn(N · Δ_t − V_{t−1})   // Update variance V
    end if
    D_t = (Δ_t ≥ V_t)                   // Compute detection image D
    if D_t == 0                         // Update background model M ...
        M_t = M_{t−1} + sgn(I_t − M_{t−1})       // with relevance feedback
    end if
end for
```

Table 1. The basic sigma-delta background estimation.

$M_t$ represents the background-model image at frame $t$, $I_t$ represents the current input image, and $V_t$ represents the temporal *variance estimator image* (or *variance image*, for short), carrying information about the variability of the intensity values at each pixel. It is used as an adaptive threshold to be compared with the difference image. Pixels with higher intensity fluctuations will be less sensitive, whereas pixels with steadier intensities will signal detection upon lower differences. The only parameter to be adjusted is $N$, with typical values between 1 and 4. Another implicit parameter in the algorithm is the updating period of the statistics, which depends on the frame rate and the number of grey levels. This updating period can be modified by performing the loop processing every $P$ frames, instead of every frame. The same algorithm computes the detection image or *detection mask*, $D_t$. This binary image highlights pixels belonging to the detected foreground objects (1-valued pixels) in contrast to the stationary background pixels (0-valued pixels). The described algorithm is, in fact, a slight variation of the basic sigma-delta algorithm, where the background model is only updated for those pixels where no detection is signalled, instead of doing it for all pixels. This selective updating is called *relevance feedback* and it is usually preferable, as it provides more stability to the background model.

### 2.2 Sigma-Delta algorithm with spatiotemporal processing

The basic sigma-delta algorithm only performs a strict temporal processing at the pixel level. Recent improvements suggest enhancing the method by adding some spatiotemporal processing (Manzanera & Richefeu, 2007). The aim of the additional spatiotemporal processing is to remove non-significant pixels from the detection mask and to reduce the "ghost" and aperture effects. The "ghost effect" is the false detection produced by an object which suddenly starts moving after a motionless stay (a slow moving vehicle causes an effect similar to a ghost-like trail which can be apparent in the background model). The aperture effect produces poor detection for those objects with weak projected motion (for instance, objects moving nearly perpendicular to the image plane). The additional processing tries to improve and regularize the achieved detection through the following three operations: common-edges hybrid reconstruction, opening by reconstruction and temporal confirmation. These operations consider several common morphological operators (Vincent, 1993; Heijmans, 1999; Salembier & Ruiz, 2002):

- $Dil_\lambda(X)$: Morphological dilation of an image X, using a ball of radius $\lambda$ as structuring element.
- $Ero_\lambda(X)$: Morphological erosion of an image X, using a ball of radius $\lambda$ as structuring element.
- $\breve{Dil}_\lambda^Y(X) = Min\,(Dil_\lambda(X), Y)$: Geodesic dilation of a marker image X, using a ball of radius $\lambda$ as structuring element and a reference image Y.
- $\breve{\mathrm{R}}\mathrm{e}c^Y(X) = \lim_{k\to\infty} X(k)$: Geodesic reconstruction of an image X (marker image), using a reference image Y. Here, the geodesic dilation is used in a recursive manner, as: $X(k) = \breve{Dil}_\lambda^Y(X(k-1))$, with $X(0) = X$. It can be shown that the series $X(k)$ defined in such a way always converges after a finite number of iterations.

Besides these classical morphological operators, a special reconstruction, called hybrid reconstruction, $\widetilde{\mathrm{R}}\mathrm{e}c_\alpha^Y(X)$, is introduced by Manzanera and Richefeu, (2007), based on the idea of gradually forgetting the marker. This operator is implemented as a four-step forgetting reconstruction, as follows:

$$
\begin{aligned}
&\widetilde{\mathrm{R}}\mathrm{e}c_\alpha^Y(X)^{(0)}(c,r) = Min\left[Y(c,r),\quad \alpha\, X(c,r) + (1-\alpha)\, Max\left(X(c,r), \widetilde{\mathrm{R}}\mathrm{e}c_\alpha^Y(X)^{(0)}(c-1,r)\right)\right]\\
&\widetilde{\mathrm{R}}\mathrm{e}c_\alpha^Y(X)^{(1)}(c,r) = Min\left[Y(c,r), \alpha\, \widetilde{\mathrm{R}}\mathrm{e}c_\alpha^Y(X)^{(0)}(c,r) + (1-\alpha)\, Max\left(\widetilde{\mathrm{R}}\mathrm{e}c_\alpha^Y(X)^{(0)}(c,r), \widetilde{\mathrm{R}}\mathrm{e}c_\alpha^Y(X)^{(1)}(c+1,r)\right)\right]\\
&\widetilde{\mathrm{R}}\mathrm{e}c_\alpha^Y(X)^{(2)}(c,r) = Min\left[Y(c,r), \alpha\, \widetilde{\mathrm{R}}\mathrm{e}c_\alpha^Y(X)^{(1)}(c,r) + (1-\alpha)\, Max\left(\widetilde{\mathrm{R}}\mathrm{e}c_\alpha^Y(X)^{(1)}(c,r), \widetilde{\mathrm{R}}\mathrm{e}c_\alpha^Y(X)^{(2)}(c,r-1)\right)\right]\\
&\widetilde{\mathrm{R}}\mathrm{e}c_\alpha^Y(X)^{(3)}(c,r) = Min\left[Y(c,r), \alpha\, \widetilde{\mathrm{R}}\mathrm{e}c_\alpha^Y(X)^{(2)}(c,r) + (1-\alpha)\, Max\left(\widetilde{\mathrm{R}}\mathrm{e}c_\alpha^Y(X)^{(2)}(c,r), \widetilde{\mathrm{R}}\mathrm{e}c_\alpha^Y(X)^{(3)}(c,r+1)\right)\right]\\
&\widetilde{\mathrm{R}}\mathrm{e}c_\alpha^Y(X) = \widetilde{\mathrm{R}}\mathrm{e}c_\alpha^Y(X)^{(3)}
\end{aligned}
\tag{1}
$$

In these expressions, $c$ and $r$ refer to the column and row of each pixel in the image, respectively, while $1/\alpha$ is the reconstruction radius replacing the structuring element.

The three operations involved in spatiotemporal processing that make use of the detailed morphological operators are then:

1. *Common-edges hybrid reconstruction:* $\Delta_t^{\nabla} = \widetilde{\mathrm{Rec}}_{\alpha}^{\Delta_t}\left(\mathrm{Min}\left(\nabla(I_t), \nabla(\Delta_t)\right)\right)$ This step tries to make a reconstruction within $\Delta_t$ of the common edges in the current image and the difference image. It is intended to reduce the eventual ghost effects appearing in the difference image. $\nabla(I)$ must be understood as the gradient module image of $I$. The minimum operator, Min(), acts like an intersection operator, but working on gray-level values, instead of binary values. This operation retains the referred common edges belonging both to $\Delta_t$ and $I_t$. Finally, the $\widetilde{\mathrm{Rec}}_{\alpha}^{\Delta_t}()$ operator performs the aforementioned reconstruction, trying to recover the whole object from its edges, but restricted to the difference image (Manzanera & Richefeu, 2007).
2. *Opening by reconstruction:* $L_t = \breve{\mathrm{Rec}}^{D_t}\left(\mathrm{Ero}_{\lambda}(D_t)\right)$. After obtaining the detection mask, this step is applied in order to remove the small connected components present in it. A binary erosion with radius $\lambda$, $\mathrm{Ero}_{\lambda}()$, followed by the usual geodesic reconstruction, restricted to $D_t$, is applied.
3. *Temporal confirmation:* $D_t^{\nabla} = \breve{\mathrm{Rec}}^{L_t}\left(L_{t-1}\right)$. The final detection mask is obtained after another reconstruction operation along time. This step, combined with the previous one, can be interpreted as: "keep the objects bigger than $\lambda$ that appear at least on two consecutive frames".

Table 2 describes the complete sigma-delta with spatiotemporal processing algorithm. Despite this rather sophisticated procedure, this algorithm also exhibits eventual problems due to its intrinsic updating period. For instance, it shows a limited adaptation capability to certain complex scenes in urban environments or, in general, scenes permanently crossed by lots of objects of very different sizes and speeds. In Manzanera and Richefeu (2007), the authors suggest overcoming this problem using the multiple-frequency sigma-delta background estimation.

```
M_0 = I_0                 // Initialize background model M
V_0 = 0                   // Initialize variance V
for each frame t
    Δ_t = |M_t − I_t|     // Compute current difference
    if Δ_t ≠ 0
        V_t = V_{t−1} + sgn(N·Δ_t − V_{t−1})   // Update variance V
    end if
    Δ_t^∇ = R̃ec_α^{Δ_t}(Min(∇(I_t), ∇(Δ_t)))   // Common-edges hybrid reconst.
    D_t = (Δ_t^∇ ≥ V_t)   // Compute initial detection mask D
    L_t = R̆ec^{D_t}(Ero_λ(D_t))   // Opening by reconstruction
    D_t^∇ = R̆ec^{L_t}(L_{t−1})    // Final det. mask after temporal confirmation
    if D_t^∇ == 0         // Update background model M ...
        M_t = M_{t−1} + sgn(I_t − M_{t−1})     // with relevance feedback
    end if
end for
```

Table 2. Sigma-delta background estimation with spatiotemporal processing.

**2.3 Multiple-frequency Sigma-Delta algorithm**

The principle of this technique is to compute a set of $K$ backgrounds $M_t^i, i \in [1,K]$, each one characterized by its own updating period $\alpha_i$. The compound background model is obtained from a weighted combination of the models in that set. Each weighting factor is directly proportional to the corresponding adaptation period and inversely proportional to the corresponding variance. The background model is improved, but at the expense of an increment in the computational cost with respect to the basic sigma-delta algorithm. Table 3 details an example of multi-frequency background estimation using $K$ different periods $\alpha_1 < \ldots < \alpha_K$.

In this case, the relevance feedback is not convenient due to fact of using several background models with different periods.

```
for each i ∈ [1,K]
    M_0^i = I_0   // Initialize background model for each period, M^i
    V_0^i = 0     // Initialize variance for each period, V^i
end for
V_0 = 0           // Initialize global variance V

for each frame t
    M_t^0 = I_t   // Initialize base-case model
    V_t^0 = 0     // Initialize base-case variance
    for each i ∈ [1,K]
        if t is a multiple of α_i
            // Recursive rule for updating background model M^i
            M_t^i = M_{t-1}^i + sgn(M_t^{i-1} - M_{t-1}^i)
        end if
        Δ_t^i = |M_t^i - I_t|   // Compute current difference with model M^i
        if Δ_t^i ≠ 0
            V_t^i = V_{t-1}^i + sgn(N · Δ_t^i - V_{t-1}^i)   // Update variance V^i
        end if
    end for
    M_t = (Σ_{i∈[1,K]} α_i M_t^i / V_t^i) / (Σ_{i∈[1,K]} α_i / V_t^i)   // Compute global background model
    Δ_t = |M_t - I_t|   // Compute current difference with global model
    if Δ_t ≠ 0
        V_t = V_{t-1} + sgn(N · Δ_t - V_{t-1})   // Update global variance
    end if
    D_t = (Δ_t ≥ V_t)   // Final detection mask
end for
```

Table 3. Multiple-frequency sigma-delta background estimation.

### 2.4 Sigma-Delta algorithm with confidence measurement

A different improvement of the basic sigma-delta background subtraction algorithm has been proposed by Toral et al., (2009b). The aim of this algorithm consists of trying to keep the high computational efficiency of the basic method, while making it particularly suitable for urban traffic environments, where very challenging conditions are common: dense traffic flow, eventual traffic congestions, or vehicle queues. In this context, background subtraction algorithms must handle the moving objects that merged into the background due to a temporary stop and then become foreground again. Many implementations overcome this problem with a subsequent post-processing or foreground validation step. The aim of this algorithm is to alleviate this subsequent step, preventing the background model to incorporate objects which are slow moving or stopped for a time gap. For this purpose, a numerical confidence level which is tied to each pixel in the current background model is introduced. This level quantifies the trust the current value of that pixel deserves. This enables a mechanism that tries to provide a better balance between adaptation to illumination or background changes in the scene and prevention against undesirable background-model contamination from slow moving vehicles or vehicles that are motionless for a time gap, without compromising the real-time implementation. The algorithm is detailed in Table 4. Three new images are required with respect to the basic sigma-delta algorithm: the frame counter image ( $I_t^{FC}$ ), the detection counter image ( $I_t^{DC}$ ) and the confidence image ( $I_t^{CON}$ ).

The variance image is intended to represent the variability of pixel intensities when no objects are over that pixel. In other words, the variance image will solely be determined by the background intensities, as a proper threshold should be chosen from that. A low variance should be interpreted as having a "stable background model" that has to be maintained. A high variance should be interpreted as "the algorithm has to look for a stable background model". One of the problems of the previous versions of sigma-delta algorithms in urban traffic environments is that, as the variance grows when vehicles are passing by, the detection degrades because the threshold becomes too high. Then, it is necessary to perform a more selective background and variance update.

The main background and variance selective updating mechanism is linked to the so-called "refresh period". Each time this period expires (let us say, each $P$ frames), the updating action is taken, provided that the traffic conditions are presumably suitable. The detection ratio can be used as an estimation of the traffic flow. Notice that this is an acceptable premise if we assume that the variance threshold filters out background intensity fluctuations, as intended. Values of this detection ratio above 80% are typically related to the presence of stopped vehicles or traffic congestion over the corresponding pixels. If this is not the case, then the updating action is permitted.

On the other hand, high variance values mean that the capability for a proper evaluation of the traffic flow is poor, as the gathered information related to the detection ratio is not reliable. In this case, it is wiser not to recommend the updating action.

A parallel mechanism is set up in order to update the confidence measurement. This second mechanism is controlled by the so-called "confidence period". This is not a constant period of time, but it depends on the confidence itself, for each particular pixel. The principle is that the higher the confidence level is, the lower the updating need for the corresponding pixel is. Specifically, the confidence period length is given by a number of frames equal to the confidence value at the corresponding pixel. Each time the confidence period expires, the

```
M_0 = I_0;  V_0 = v_ini        // Initialize background model and variance
I_0^DC = I_0^FC = 0;  I_0^CON = c_ini   // Initialize detection, frame counter and confidence measure
for each frame t
  I_t^FC = I_t^FC + 1        // Increment frame-counter image
  // Period evaluation and background updating decision making:
  if I_t^FC < I_t^CON        // If current confidence period not expired yet
    if I_t^FC is a multiple of P  // If refresh period expires
      if V_t ≤ v_th  //Low variance => we assume we can rely on the gathered information (in
                         particular in the detection counter) => traffic flow may be evaluated
        if (I_t^DC / I_t^FC) ≤ 0.8   // If not very heavy traffic
          U_t = 1                    // Refresh period updating mode
        end if
      end if
    end if
  else    // If current confidence period expires
    if V_t ≤ v_th  // Low variance => we assume we can evaluate traffic flow
      I_t^CON += γ(I_t^DC / I_t^FC)  // Confidence updating as a function of the detection ratio
      if I_t^CON == c_min    // If confidence goes down to the minimum …
        U_t = 1              // … force updating
      end if
    else                 // We cannot reliably evaluate traffic flow
      U_t = 1   // Confidence period updating mode, to avoid background model deadlock
    end if
    I_t^DC = I_t^FC = 0  // Reset detection counter and frame counter
  end if
  // Background updating (if appropriate) and detection:
  if U_t == 1  // If updating recommended, follow sigma-delta algorithm
    M_t = M_{t-1} + sgn(I_t - M_{t-1})   // Update background model
    Δ_t = |M_t - I_t|                    // Compute current difference
    V_t = V_{t-1} + sgn(v_min + N·Δ_t - V_{t-1})  // Update variance
    D_t = (Δ_t ≥ V_t)        // Compute detection mask
  else                       // Do not update, just detect
    Δ_t = |M_t - I_t|
    D_t = (Δ_t ≥ V_t)
  end if
  I_t^DC += (D_t == 1)         // Update detection-counter image
end for
```

Table 4. Sigma-delta algorithm with confidence measurement.

confidence measure is incrementally updated, according to an exponentially decreasing function of the detection ratio, $d$:

$$\gamma(d) = round(\alpha \cdot \exp(-\beta d) - 1) \tag{2}$$

The gain $\alpha$ is tuned as the confidence maximum increment (when the detection ratio tends to zero), while $\beta$, defining the increment decay rate, has to be chosen such that negative increments are restricted to large detection rates.

The recommended values are, $\alpha$ = 11, so the maximum confidence increment is 10 frames, and $\beta$ = 4 which adjusts the crossing of the function with -0.5 around 75%-80% of detection rates.

In case the confidence is decremented down to a minimum, background updating is forced. This is a necessary working rule since, in the case of cluttered scenes, for instance, the background model may not be updated by means of the refresh period. Thus, in that case, this underlying updating mechanism tries to prevent the model to get indefinitely locked in a wrong or obsolete background.

As a last resort, there is another context in which the updating action is commanded. This is the case when the confidence period expires but the detection capability is estimated to be poor. In such a case, as no reliable information is available, it is preferred to perform the background update. In fact, by doing otherwise, we will never change the situation, as the variance won't be updated, hence the algorithm would end in a deadlock.

The confidence measurement is related to the maximum updating period. In very adverse traffic conditions, this period is related to the time the background model is able to keep untainted from the foreground objects. Let us suppose a pixel with correct background intensity and maximum confidence value, for instance, $c_{max}$ = 125 frames. Then, 125 frames have to roll by for the confidence period to expire. If the traffic conditions do not get better, the confidence measure decreases until 124 and no updating action is taken. Now, 124 frames have to roll by for the new confidence period to expire. At the end, 125+124+123+…+10 = 7830 frames are needed for the algorithm to force the updating action (assuming minimum confidence value, $c_{min}$ = 10). At the typical video rate of 25 frames per second, this corresponds to more than 5 minutes before the background starts becoming corrupted if the true background is seldom visible due to a high-traffic density. The downside is that, if we have a maximum confidence for a pixel with wrong intensity (for instance, if the background of the scene itself has experienced an abrupt change), also this same period is required for the pixel to be adapted to the new background. Nevertheless, if the change in the background is a significant illumination change, this problem can be alleviated in a further step by employing techniques related to shadow removal, which is beyond the scope of this paper (Prati et al., 2003; Cucchiara et al., 2003).

When the evaluation of the confidence measurement and the detection ratio recommend taking the updating action, the basic sigma-delta algorithm is applied. If no updating is required, the computation of the detection mask is just performed.

## 3. Comparative results

### 3.1 Qualitative performance analysis

A typical traffic urban sequence is used in this qualitative comparative study. In such scenes the background model from the basic sigma-delta algorithm quickly degrades, assimilating the slow moving or stopped vehicles. Another undesirable effect is that, in the long term, the corresponding variance values tend to increase immoderately in the areas with a higher traffic density. As the variance is used as a detection threshold, this detection is not very sensitive, producing a poor detection mask. This is illustrated in Fig. 1 for the *traffic-light sequence*. The first column of the figure shows the current image at frame 400 (that is 16 seconds after the sequence starts), which is the same for every row. The second column represents the current background model for each compared method. The third column represents the visual appearance of the variance image, and the fourth column represents the detection mask. The results shown in the first row corresponds to the basic sigma-delta algorithm, *SD* (parameter settings: *N*=4). The second row corresponds to the sigma-delta with spatiotemporal processing, $SD^{SP}$ (parameter settings: *N*=4, $\alpha = 1/8$, $\lambda = 1$), while the third row represents the results from the multiple-frequency sigma-delta background estimation, $SD^{M}$ (*N*=4, *K*=3 backgrounds models used, with adaptation periods: $\alpha_1 = 1$, $\alpha_2 = 8$ and $\alpha_3 = 16$). Finally, the fourth row corresponds to the sigma-delta with confidence measurement, *SDC* (parameter settings: *N*=4, $V_t \in [v_{\min}, v_{\max}] = [10, 200]$, $[c_{\min}, c_{\max}] = [10, 125]$, $v_{ini} = v_{\min}$, $c_{ini} = c_{\min}$, $P = c_{\min}$, $v_{th} = 38$).

It can be seen that the adaptation speed of multi-frequency sigma-delta and the proposed method (when it is seeking for a new background) are similar. In particular, the moving vehicles present in the image at the beginning of the sequence have not been completely "forgotten" yet, producing the ghost vehicles noticeable in the case of these two algorithms. On the other hand, we can appreciate the effect of ghostly trails apparent in the background model, produced by the slow moving vehicles (or vehicles moving in a direction nearly perpendicular to the image plane), in the case of the two first algorithms.

Fig. 2 illustrates another sample of the behaviour of these four algorithms at frame 1200 (48 seconds after the sequence start). In this case, some vehicles have been stopped in front of a red light for a maximum of 20 seconds approximately. It can be seen that these vehicles have been blended into the background model for both, the basic sigma-delta and sigma-delta with spatiotemporal processing, while they have been partially blended into the background for the multi-frequency sigma-delta. The sigma-delta with confidence measurement algorithm keeps this background model unpolluted from those stopped vehicles, being able to attain its full detection as foreground items. It can also be observed that the variance values have not been significantly increased in the region of the stopped vehicles, keeping the detection threshold conveniently sensitive.

Next, the situation a few seconds later is shown in Fig. 3, corresponding to frame 2170, 86 seconds after the beginning of the sequence, and around 15 seconds after the vehicles in front of the traffic light started moving again. It can be seen that those vehicles have not been completely "forgotten" from the background model in the case of the basic sigma-delta, the sigma-delta with spatiotemporal processing and the multi-frequency sigma-delta algorithms. On the other hand, since this frame has been preceded by a significant traffic flow, the variance in the case of the first three algorithms has raised accordingly, producing a poor detection in the areas with higher variance. On the contrary, the sigma-delta with confidence measurement algorithm tries to keep the variance conveniently sensitive in those areas, as the variance is intended to represent the variability of the intensity levels of the background pixels only.

Finally, in Fig. 4, the situation 240 seconds after the sequence start is shown. This frame is part of the third red light cycle. The same comments made with respect to Fig. 2 are extensible to this later fragment of the sequence.

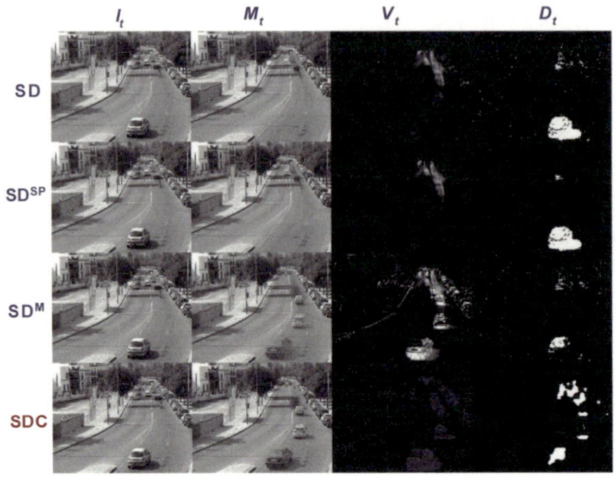


Fig. 1. Traffic-light sequence. Comparative results at frame 400.

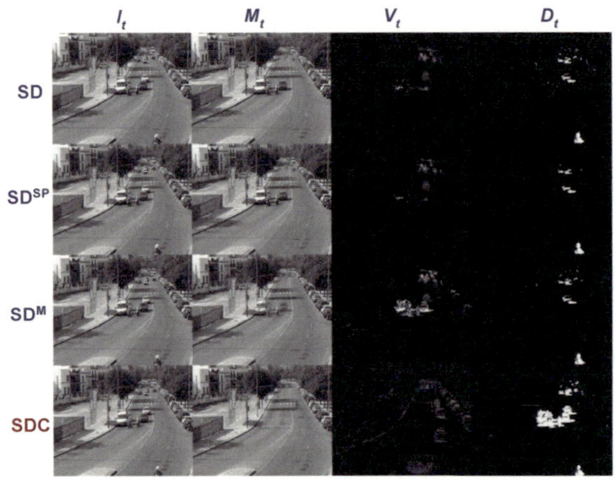


Fig. 2. Traffic-light sequence. Comparative results at frame 1200.

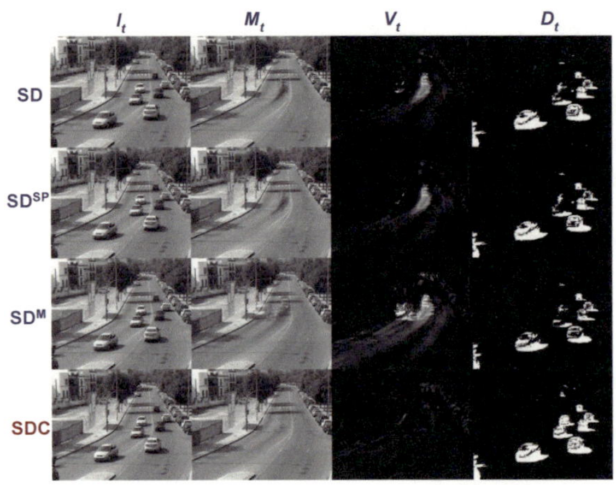


Fig. 3. Traffic-light sequence. Comparative results at frame 2170.

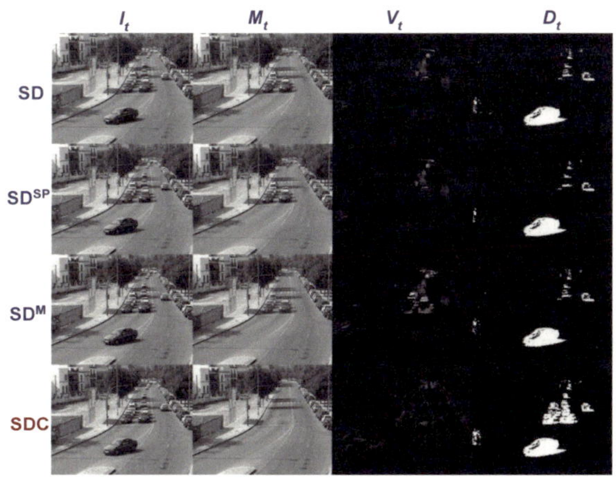


Fig. 4. Traffic-light sequence. Comparative results at frame 6000.

### 3.2 Quantitative performance analysis

There are different approaches to evaluate the performance of the background subtraction algorithms, from low-level, pixel-oriented evaluation to object-level or application-level evaluation. In the latter case, the goal-based evaluation of the foreground detection would be influenced by other higher level components of the application, e.g. a blob feature extraction module or a tracker module, which are out the scope of this paper. Consequently, in this section, a pixel-oriented evaluation has been preferred.

In a binary decision problem, the classifier labels samples as either positive or negative. In our context, samples are pixel values, "positive" means foreground object pixel, and "negative" means background pixel. In order to quantify the classification performance, with respect to some ground truth classification, the following basic measures can be used:

- True positives (TP): correctly classified foreground pixels.
- True negatives (TN): correctly classified background pixels.
- False positives (FP): incorrectly classified foreground pixels.
- False negatives (FN): incorrectly classified background pixels.

Using these basic measures, the true and false positive rates can be estimated:

$$\text{True positive rate: } TPR = \frac{TP}{total\ of\ actual\ positives} = \frac{TP}{TP+FN} \tag{3}$$

$$\text{False positive rate: } FPR = \frac{FP}{total\ of\ actual\ negatives} = \frac{FP}{TN+FP} \tag{4}$$

Precision and recall are defined as:

$$\text{Precision: } PR = \frac{TP}{total\ of\ estimated\ positives} = \frac{TP}{TP+FP} \tag{5}$$

$$\text{Recall: } RE = TPR \tag{6}$$

Other measures for fitness quantification, in the context of background subtraction techniques, have been proposed in the literature (Rosin & Ioannidis, 2003; White & Shah, 2007; Ilyas et al., 2009). The following are some examples:

$$\text{F-measure: } S_F = 2\left(\frac{PR \cdot RE}{PR + RE}\right), \qquad (0 \le S_F \le 1) \tag{7}$$

which combines precision and recall in the form of their harmonic mean, providing an index more representative than the pure *PR* and *RE* measures themselves.

$$\text{Percentage of correct classification: } S_{CC} = \frac{TP + TN}{TP + TN + FP + FN}, \qquad (0 \le S_{CC} \le 1) \tag{8}$$

The percentage of correct classification alone is very commonly used for assessing a classifier's performance. However, it can give misleading estimates when there is a significant skew in the class distribution (Rosin & Ioannidis, 2003). In particular, if foreground elements are only present in a small part of the image, lets say 5%, there is not much difference in the achieved high ratings of this coefficient with respect to the case of simply classifying everything as background. Using additionally the Jaccard and Yule coefficients (Sneath & Sokal, 1973) can reduce the problem, when there is a large volume of expected true negatives:

$$\text{Jaccard coefficient: } S_J = \frac{TP}{TP + FP + FN}, \qquad 0 \le S_J \le 1 \tag{9}$$

$$\text{Yule coefficient: } S_Y = \frac{TP}{TP + FP} + \frac{TN}{TN + FN} - 1 = PR + PR_N - 1, \qquad (-1 \le S_Y \le 1) \tag{10}$$

$PR_N$ has to be understood as the precision in the background classification (negatives), in the same way *PR* is the precision in the foreground classification (positives). In its original form, the Yule coefficient is defined on the interval [-1,1]. The lower limit of this interval occurs when there are not matching pixels, while a perfect match would make the coefficient to hit the upper bound.

Finally, Ilyas et al. (2009) proposes a weighted Euclidean distance, considering the deviations of *FPR* and *TPR* from their respective ideal values, 0 and 1. It is defined as follows:

$$E_\gamma = \sqrt{\gamma\, FPR^2 + (1-\gamma)\,(1-TPR)^2}, \qquad (0 \le E_\gamma \le 1) \tag{11}$$

where $\gamma$ (0< $\gamma$ <1) is a weighting coefficient, that has to be adjusted according to the desired trade off between sensitivity and specificity. For instance, when a low false alarm rate is the priority, at the expense of loosing sensitivity, high values for this coefficient have to be chosen.

A representative ground truth dataset has been elaborated using the traffic light sequence. A number of samples from the traffic light sequence have been extracted and manually annotated using the publicly available annotation tool: *InteractLabeler* (Brostow et al., 2009). One ground-truth frame for every 100 frames has been picked out, which corresponds to a 0.25 fps sampling rate. An initialization stage of around 20 seconds long is skipped over.

The following index sets have been considered as a valuable quantification of relative performance of each algorithm: $S = \{S_F,\ S_J,\ 0.5(1+S_Y)\}$, $E = \{E_{0.25},\ E_{0.50},\ E_{0.75})\}$. The first set includes *fitness coefficients* with ideal value equal to 1, while the second set includes *fitness errors* with ideal value equal to 0. The values of each one of these coefficients will be averaged for all the chosen frames. In addition, the typical deviation of each one is calculated. It can be noticed that the $S_{CC}$ coefficient has been dropped from the analysis as, in agreement with the above comments, it exhibited a very poor sensitivity, yielding very high scores for every algorithm.

Table 5 details the average, $\mu$, and typical deviation, $\sigma$, for the chosen fitness indexes and the traffic light video sequence. According to the authors in Manzanera & Richefeu (2007), parameter $N$=2 was recommended for the *SD*, $SD^{SP}$ and $SD^{M}$ algorithms. However, in our experiments $N$=4 performed slightly better in some videos. Therefore, both values have been considered for each one of the four sigma-delta alternatives.

Results of the proposed sigma-delta algorithm with confidence measurement are clearly on top according to the fitness coefficients, both for $N$=2 and $N$=4. On the other hand, according to the fitness error coefficients, the proposed algorithm is significantly better than any of the other algorithms, featuring also a moderate typical deviation.

| Algorithm | $S_F$ | | $S_J$ | | 0.5 (1+$S_Y$) | | $E_{0.25}$ | | $E_{0.50}$ | | $E_{0.75}$ | |
|---|---|---|---|---|---|---|---|---|---|---|---|---|
| | $\mu$ | $\sigma$ | $\mu$ | $\sigma$ | $\mu$ | $\sigma$ | $\mu$ | $\sigma$ | $\mu$ | $\sigma$ | $\mu$ | $\sigma$ |
| **SD** (N=2) | 0,44 | 0,156 | 0,29 | 0,125 | 0,66 | 0,072 | 0,36 | 0,143 | 0,30 | 0,115 | 0,22 | 0,078 |
| **SD** (N=4) | 0,59 | 0,172 | 0,44 | 0,161 | 0,87 | 0,062 | 0,45 | 0,152 | 0,36 | 0,124 | 0,26 | 0,087 |
| **$SD^{SP}$** (N=2) | 0,63 | 0,180 | 0,48 | 0,177 | 0,92 | 0,027 | 0,42 | 0,168 | 0,34 | 0,137 | 0,24 | 0,097 |
| **$SD^{SP}$** (N=4) | 0,56 | 0,185 | 0,41 | 0,167 | 0,94 | 0,018 | 0,50 | 0,151 | 0,41 | 0,123 | 0,29 | 0,087 |
| **$SD^{M}$** (N=2) | 0,42 | 0,160 | 0,28 | 0,126 | 0,66 | 0,075 | 0,40 | 0,141 | 0,33 | 0,114 | 0,24 | 0,078 |
| **$SD^{M}$** (N=4) | 0,55 | 0,169 | 0,40 | 0,153 | 0,87 | 0,064 | 0,49 | 0,142 | 0,40 | 0,116 | 0,28 | 0,082 |
| **SDC** (N=2) | 0,83 | 0,031 | 0,72 | 0,045 | 0,92 | 0,025 | 0,16 | 0,032 | 0,13 | 0,026 | 0,09 | 0,018 |
| **SDC** (N=4) | 0,83 | 0,034 | 0,70 | 0,048 | 0,95 | 0,016 | 0,21 | 0,044 | 0,17 | 0,036 | 0,12 | 0,025 |

Table 5. Quantitative evaluation of the foreground segmentation for the traffic-light sequence.

## 4. Hardware implementation

Embedded multimedia processors are expected to be on the basis of future ITS electronic equipments (Barrero et al., 2010). From a hardware point of view, a RISC processor is the core component of the multimedia processor. They can easily support an operating system for managing interfaces to communication channels like Ethernet and wireless devices, to massive storage devices like MMC/SD cards and USB pen drives, and to digital I/O or LCD screens. The multimedia processor also includes special support for video applications.

From a software point of view, embedded processors can incorporate not only computational intensive algorithms to allow traffic parameter estimation or incident detection, but also standard communication protocols, interface to data storage media (USB,

MMC) and wireless connectivity (Bluetooth), FTP and SSH servers for software upgrading and a web server for remote configuration. Because of these complexities, it is advisable the use of an embedded operating system allowing the application programmers to focus on higher-level functionalities, like those based on artificial vision techniques. The selected operating system should support preemptive multitasking or multi-threading and device drivers for required connectivity. In this context, RISC processors are especially well suited for running such an operating system, offering a large spectrum of choices, both open source or proprietary.

A prototype based on the *Freescale i.MX21* multimedia processor has been developed to deal with the computer vision functionality as well as the multimedia and networking capabilities (Fig. 5). The i.MX21 processor includes an ARM926EJ-S core, operating at 266 MHz. On-chip modules include LCD and memory card/secure digital controllers (MMC/SD), serial port controller (USB OTG), CMOS sensor interface, and enhanced MultiMedia Accelerator (eMMA), which consists of a video Pre-processor (PrP) unit, an Encoder (ENC) and a Decoder (DEC) module, and a Post-processor (PP) unit. MPEG-4 and H.263 protocols are supported as well as real-time encoding/decoding of images from 32×32 pixels up to CIF format at 30 frames per second. The PrP resizes input frames from memory or from the CMOS sensor interface, and performs color space conversion. The PP module takes raw images from memory and performs additional processing to de-block, de-ring, resize, and color space conversion on the decoded frames for MPEG-4 video streaming. The prototype board is also provided with mini USB and SD card connector, RJ45 Ethernet connector, Bluetooth expansion connector, and a coaxial connector, which provides the analog signal to a video decoder chip.

Fig. 5. Hardware prototype.

Although a multitude of embedded operating systems are currently available (Wind River's VxWorks, Microsoft Windows CE, QNX Neutrino, etc), Linux is firmly in first place as the operating system of choice for smart gadgets and embedded systems (Toral et al., 2009c). All embedded operating systems require a considerable effort of customization, because they incorporate a wider variety of input/output devices than typical desktop computers. As a consequence, it is necessary to adapt the operating system to the particular features of the selected processor. Fortunately, Linux comes under a GPL license and the community of support around Linux ports are of great help during the customization task. Linux runs on multiple embedded architectures, but ARM and PowerPC are the best supported processors. Besides, Linux support multitasking/multithreading, allowing several processes and services running in concurrent operation. In the proposed application, the i.MX21 video processor is running under ARM Linux (kernel 2.4.20). The main process corresponds to the background model estimation, but several processes executing additional services are in concurrent operation:

- A HTTP server for configuration and supervision purposes.
- A Web command server module, in charge of processing specific requests from the configuration web page.
- A SSH server for remote logging into the system.
- A FTP server for upload and download operations.
- A watchdog process for rebooting the system when a periodic signal is not received from the main process.
- A video delivery process for video compression and delivery using an Ethernet interface.

All the sigma-delta algorithms have been programmed using C++ programming language. Full resolution, gray-scale images 720x576 pixels are subsampled to resolution 360×288 before being processed. Table 6 shows the time requirements of each one of the considered algorithms, while performing on a typical traffic sequence.

Considering the average time or the effective velocity columns, the basic sigma-delta algorithm is the less time consuming, as expected, followed by the proposed algorithm. The multifrequency sigma-delta takes around 2.5 times more than the proposed algorithm, while the sigma-delta with spatio-temporal processing is about 13 times slower. On the other hand, regarding to the maximum cycle time, we can see that both, the multifrequency sigma-delta and the SDC algorithm, double their respective average times, while the sigma-delta with spatio-temporal processing triplicate its average time in the worst case. Furthermore, the basic sigma-delta algorithm and the enhanced version with confidence measurement have the benefit of a much lower typical deviation in its cycle time.

| Algorithm | | Tmin (ms) | Tmax (ms) | Tmean (ms) | Tdesv (ms) | Speed (fps) |
|---|---|---|---|---|---|---|
| SD | (N=4) | < 1 | 31 | 11,37 | 6,84 | **87,95** |
| $SD^{SP}$ | (N=4) | 218 | 1201 | 387,07 | 52,48 | **2,54** |
| $SD^{M}$ | (N=4) | 31 | 172 | 77,64 | 14,04 | **12,88** |
| SDC | (N=4) | 15 | 62 | 30,18 | 5,71 | **33,14** |

Table 6. Time requirements of each algorithm.

## 5. Applications

The main process of the prototype corresponds to vision-based vehicle detection system. This detection relies on the so called detection areas. Each one of these areas or regions is a user-configurable polygon with an arbitrary shape or size, and an associated functionality. Three kinds of detection regions have been programmed: presence, directional and queue regions, allowing the estimation of useful traffic information. The functionality of these regions is clarified below:

- Presence-detection regions. They may be considered as virtual loop detectors, quite similar to the traditional on-the-road loop detectors buried under the road surface (Michalopoulos, 1991). These non-invasive loop detectors generate a binary output (vehicle presence or absence) depending on a configurable threshold, and incorporate a software vehicle counting functionality.

- Directional-detection regions. The directional regions are also used for counting vehicles. Unlike the previous type, its goal is to estimate the vehicle running direction, checking how close it is to the configured direction associated to the region. Depending on the agreement of both directions, the vehicle is counted or ignored. This kind of regions is useful for selective vehicle counting in or near intersections, one-way violation detections or restricted turn infringements.
- Queue-detection regions. The queue regions are intended to estimate vehicle queue length and queuing frequency, typically, in front of a traffic light. A binary output can be also associated with these regions, indicating whether the instantaneous queue length is above or below a configured threshold.

This main process requires the complete configuration of the scene, which can be made via the HTTP server running in the multimedia processor. Fig. 6 illustrates one of the screens in the system's configuration web page.

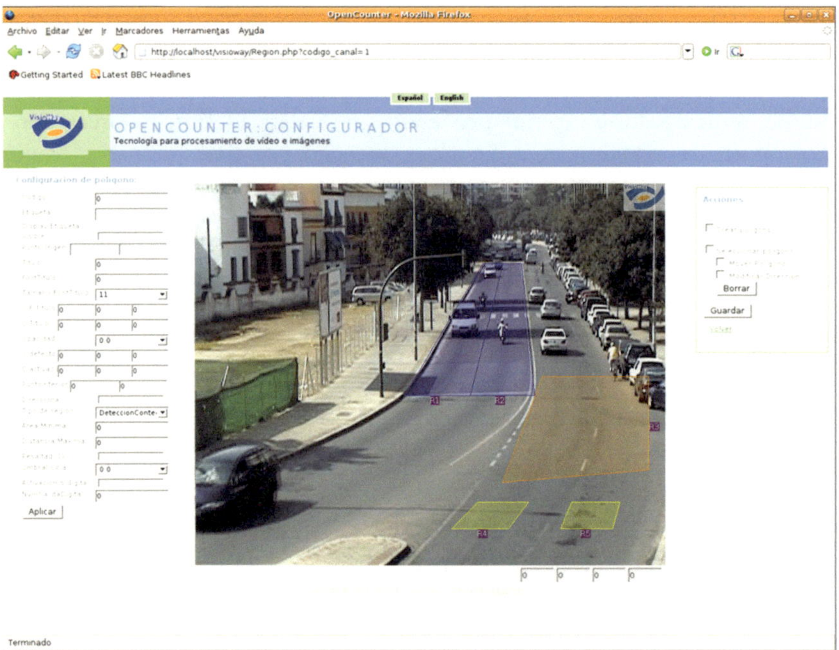


Fig. 6. Equipment configuration web site.

The image on the centre of the web page shows a snapshot of a video stream delivered by the real-time application. Several detection regions have been defined: two presence-detection regions in yellow configured to count vehicles running on both lanes, one directional detection region in pink configured to count vehicles coming from the right side of the intersection, and two queue regions in blue in front of a traffic light. Associated to each detection area, both instantaneous and time-aggregated data can be obtained:

- Presence-detection regions: occupancy and vehicle counting data during the aggregation period. The occupancy gives an estimation of the percentage of time the presence level is above the configured threshold (presence on).

- Directional-detection regions: directional vehicle counting data.
- Queue-detection regions: queue counting data (number of times the queue has exceeded the configured threshold during the aggregation period), time ratio the queue is above the configured threshold (queue on) and average queue length.

Instantaneous data overlaid on the real-time image (upper left corner of the image of Fig. 6), and produce digital outputs emulating traditional detectors. The aggregated data can be also recorded on a log file (which is updated at the end of each aggregation period), and then downloaded using the FTP server.

## 6. Conclusion

In this chapter, several background modelling techniques have been described, analyzed and tested. In particular, different algorithms based on sigma-delta filter have been considered due to their suitability for embedded systems, where computational limitations affect a real-time implementation. A qualitative and a quantitative comparison have been performed among the different algorithms. Obtained results show that the sigma-delta algorithm with confidence measurement exhibits the best performance in terms of adaptation to particular specificities of urban traffic scenes and in terms of computational requirements. A prototype based on an ARM processor has been implemented to test the different versions of the sigma-delta algorithm and to illustrate several applications related to vehicle traffic monitoring and implementation details.

## 7. References

Barrero, F.; Toral, S.; Vargas, M. & Cortés, F. (2010). Internet in the Development of Future Road-Traffic Control Systems. Internet Research, Vol. 20, Iss. 2, (Apr. 2010), pp. 154-168, ISSN 1066-2243.

Beymer, D. & Malik, J. (1996). Tracking vehicles in congested traffic. *Proceedings of the 1996 IEEE Intelligent Vehicles Symposium*, pp. 130-135, ISBN 0-7803-3652-6, Tokyo, Japan, Sept. 1996, IEEE, NJ, USA.

Beymer, D.; Mclauchlan, P.; Coifman, B. & Malik, J. (1997). A real-time computer vision system for measuring traffic parameters. *Proc. IEEE Conference on Computer Vision and Pattern Recognition*, pp. 495-501, ISBN 0-8186-7822-4, San Juan, Puerto Rico, June 1997, IEEE Computer Society, CA, USA.

Brostow, G.J.; Fauqueur, J. & Cipolla, R. (2009). Semantic object classes in video: A high-definition ground truth database. *Pattern Recognition Letters*, Vol. 30, Iss. 2, (Jan. 2009), pp. 88-97, ISSN 0167-8655.

Chalidabhongse, T.H.; Kim, K.; Harwood, D. & Davis, L. (2003). A perturbation method for evaluating background subtraction algorithms. *In Proc. 2003 Joint IEEE Int. Workshop on Visual Surveillance and Performance Evaluation of Tracking and Surveillance*, pp. 1-7, Nice, France, Oct. 2003.

Cheung, S.-C. & Kamath, C. (2004). Robust techniques for background subtraction in urban traffic video. In *Proc. of Video Communications and Image Processing*, SPIE Electronic Imaging, Vol. 5308, pp. 1-12, ISBN 9780819452115, Jan. 2004, SPIE, San Jose, California, USA.

Coifman, B.; Beymer, D.; McLauchlan, P. & Malik, J. (1998). A real-time computer vision system for vehicle tracking and traffic surveillance. *Transportation Research Part C: Emerging Technologies*, Vol. 6, Iss. 4, (Aug. 1998), pp. 271-288, ISSN 0968-090X.

Cucchiara, R.; Grana, C.; Piccardi, M. & Prati, A. (2003). Detecting moving objects, ghosts, and shadows in video streams. *IEEE Transactions on Pattern Analysis and Machine Intelligence*, Vol. 25, No. 10, (Sept. 2003), pp. 1337-1342, ISSN 0162-8828.

Elgammal, A.; Duraiswami, R.; Harwood, D. & Davis, L.S. (2002). Background and foreground modeling using nonparametric kernel density estimation for visual surveillance. *Proceedings of the IEEE*, Vol. 90, Iss. 7, (July 2002), pp. 1151-1163, ISSN 0018-9219.

Fathy, M. & Siyal, M. (1998). A window-based image processing technique for quantitative and qualitative analysis of road traffic parameters. *IEEE Transactions on Vehicular Technology*, Vol. 47, Iss. 4, (Nov. 1998), pp. 1342-1349, ISSN 0018-9545.

Ha, D.M.; Lee, J.-M. & Kim, Y.-D. (2004). Neural-edge-based vehicle detection and traffic parameter extraction. *Image and Vision Computing*, Vol. 22, No. 11, (Sept. 2004), pp. 899–907, ISSN 0262-8856.

Haritaoglu, I.; Harwood, D. & Davis, L. (2000). W4: real-time surveillance of people and their activities. *IEEE Transactions on Pattern Analysis and Machine Intelligence*, Vol. 22, Iss. 8, (Aug. 2000), pp. 809-830, ISSN 0162-8828.

Harville, M. (2002). A framework for high-level feedback to adaptive, per-pixel, mixture-of-Gaussian background models. *Lecture Notes in Computer Science*, Vol. 2352, (Jan. 2002), pp. 37-49, ISSN 0302-9743.

Heijmans, H.J.A.M. (1999). Connected Morphological Operators for Binary Images. *Computer Vision and Image Understanding*, Vol. 73, Iss. 1, (Jan. 1999), pp. 99-120, ISSN: 1077-3142.

Ilyas, A.; Scuturici, M. & Miguet, S. (2009). Real time foreground-background segmentation using a modified Codebook model. In *Proc. of the Sixth IEEE Int. Conference on Advanced Video and Signal Based Surveillance (AVSS'09)*, pp.454-459, ISBN 978-1-4244-4755-8, Genoa, Italy, Sept. 2009, IEEE Computer Society, CA, USA.

Kanhere, N.K. & Birchfield, S.T. (2008). Real-time incremental segmentation and tracking of vehicles at low camera angles using stable features. *IEEE Transactions on Intelligent Transportation Systems*, Vol. 9, Iss. 1, (Feb. 2008), pp. 148-160, ISSN 1524-9050.

Kanhere, N.K. & Birchfield, S.T. (2008). Real-time incremental segmentation and tracking of vehicles at low camera angles using stable features. *IEEE Transactions on Intelligent Transportation Systems*, Vol. 9, Iss. 1, (March 2008), pp. 148-160, ISSN 1524-9050.

Karmann, K.-P. & Brandt, A. (1990). Moving object recognition using an adaptive background memory. In *Time-Varying Image Processing and Moving Object Recognition*, V. Cappellini (Editor), Vol. 2, pp. 289-307, Elsevier Science Publishers B.V., ISBN 0444823077, Amsterdam, The Netherlands.

Kim, K.; Chalidabhongse, T.H.; Harwood, D. & Davis, L. (2005). Real-time foreground-background segmentation using codebook model. *Real-Time Imaging*, Vol. 11, No. 3, (June 2005), pp. 172-185, 2005, ISSN 1077-2014.

Koller, D.; Weber, J.; Huang, T.; Malik, J.; Ogasawara, G.; Rao, B. & Russell, S. (1994). Towards robust automatic traffic scene analysis in real-time. *Proceedings of the 12th IAPR International Conference on Computer Vision and Image Processing*, ICPR'94, Vol. 1, pp 126-131, ISBN 0-8186-6265-4, Jerusalem, Israel, Oct. 1994, IEEE Computer Society, CA, USA.

Koller, D.; Weber, J. & Malik, J. (1994). Robust multiple car tracking with occlusion reasoning, *Lecture Notes in Computer Science*, Vol. 800, (May 1994), pp. 189-196, ISSN 0302-9743.

Lo, B.P.L. & Velastin, S.A. (2001). Automatic congestion detection system for underground platforms. *Proceedings of 2001 International Symposium on Intelligent Multimedia, Video and Speech Processing*, pp. 158-161, ISBN: 962-85766-2-3, Hong Kong, China, May 2001, IEEE, NJ, USA.

Manzanera, A. & Richefeu, J. C. (2004). A robust and computationally efficient motion detection algorithm based on sigma-delta background estimation. *In Proc. of the Indian Conf. on Computer Vision, Graphics and Image Processing, ICVGIP'04*, pp. 46-51, ISBN 81-7764-707-5, Kolkata, India, Dec. 2004, Allied Publishers Private Limited, India.

Manzanera, A., & Richefeu, J. C. (2007). A new motion detection algorithm based on Sigma-Delta background estimation. *Pattern Recognition Letters*, Vol. 28, Iss. 3, (Feb. 2007). pp. 320-328, ISSN 0167-8655.

Manzanera, A. (2007). $\Sigma$ - $\Delta$ Background Subtraction and the Zipf Law. *Lecture Notes in Computer Science*, Vol. 4756, (Nov. 2007), pp. 42-51, ISSN 0302-9743.

McFarlane, N. & Schofield, C. (1995). Segmentation and tracking of pigglets in images. *Machine Vision and Applications*, Vol. 8, no. 3, (May 1995), pp. 187-193, ISSN 0932-8092.

Michalopoulos, P.G. (1991). Vehicle detection video through image processing: the Autoscope system. *IEEE Transactions on Vehicular Technology*, Vol. 40, Iss. 1/2, (Feb. 1991), pp. 21-29, ISSN 0018-9545.

Mittal, A. & Paragios, N. (2004). Motion-based background subtraction using adaptive kernel density estimation. *In Proc. of the 2004 IEEE Conference in Computer Vision and Pattern Recognition*, pp. 302-309, ISBN 0-7695-2158-4, Washington, USA, June 2004, IEEE Computer Society, CA, USA.

Monnet, A.; Mittal, A.; Paragios, A. & Ramesh, V. (2003). Background modelling and subtraction of dynamic scenes. *The Proceedings of the Ninth IEEE International Conference on Computer Vision*, pp. 1305-1312, ISBN 0-7695-1950-4, Nice, France, Oct. 2003, IEEE Computer Society, CA, USA.

Oliver, N.M.; Rosario, B. & Pentland, A.P. (2000). A Bayesian computer vision system for modeling human interactions. *IEEE Transactions on Pattern Analysis and Machine Intelligence*, Vol. 22, Iss. 8, (Aug. 2000), pp. 831-843, ISSN 0162-8828.

Piccardi, M. (2004). Background subtraction techniques: a review. *Proc. 2004 IEEE International Conference on Systems, Man and Cybernetics*, Vol. 4, pp. 3099-3104, ISBN 0-7803-8566-7, The Hague, Netherlands, Oct. 2004, IEEE, NJ, USA.

Power, W. & Schoonees, J.A. (2002). Understanding background mixture models for foreground segmentation. *Proc. of 2002 Image and Vision Computing*, pp. 267-271, Auckland, New Zealand, Nov. 2002.

Prati, A., Mikic, I., Trivedi, M.M., Cucchiara, R. (2003). Detecting moving shadows: algorithms and evaluation. *IEEE Transactions on Pattern Analysis and Machine Intelligence*, Vol. 25, No. 7, (July 2003), pp. 918-923, ISSN 0162-8828.

Rosin, P.L. & Ioannidis, E. (2003). Evaluation of global image thresholding for change detection. *Pattern Recognition Le*tters, Vol. 24, Iss. 14, (Oct. 2003), pp. 2345-2356, ISSN 0167-8655.

Salembier, P. & Ruiz, J. (2002). Connected operators based on reconstruction process for size and motion simplification. In *Proc. of IEEE Int. Conference on Acoustics, Speech, and Signal Processing, ICASSP'02*, Vol. 4, pp. 3289–3292, ISBN: 0-7803-7402-9, Orlando, Florida, USA, May 2002, IEEE Signal Processing Society, NJ, USA.

Sneath, P. & Sokal, R. (1973). *Numerical taxonomy. The principle and practice of numerical classification*. W.H. Freeman, ISBN-10 0716706970, San Francisco.

Spagnolo, P.; Orazio, T.D.; Leo, M. & Distante, A. (2006). Moving object segmentation by background subtraction and temporal analysis. *Image and Vision Computing*, Vol. 24, No. 5, (May 2006), pp. 411–423, ISSN 0262-8856.

Stauffer, C. & Grimson, W. (1999). Adaptive background mixture models for real-time tracking. *Proc. 1999 IEEE Conference on Computer Vision and Pattern Recognition*, Vol. 2, pp. 246-252, ISBN 0-7695-0149-4, Fort Collins, Colorado, USA, June 1999, IEEE Computer Society, CA, USA.

Stauffer, C. & Grimson, W. (2000). Learning patterns of activity using real-time tracking. *IEEE Transactions on Pattern Analysis and Machine Intelligence*, Vol. 22, No. 8, (Aug. 2000), pp. 747-757, ISSN 0162-8828.

Toral, S.L.; Vargas, M. & Barrero, F. (2009a). Embedded multimedia processors for road-traffic parameter estimation", *Computer*, Vol. 42, Iss. 12, (Dec. 2009), pp. 61-68, ISSN 0018-9162.

Toral, S.L.; Vargas, M.; Barrero, F. & Ortega, M.G. (2009b). Improved Sigma-Delta Background Estimation for Vehicle Detection. *Electronics Letters*, Vol. 45, Iss. 1, (Jan. 2009), pp. 32-34, ISSN 0013-5194.

Toral, S.L.; Matínez-Torres, M.R. & Barrero, F. (2009c). Virtual Communities as a resource for the development of OSS projects: the case of Linux ports to embedded processors. *Behaviour and Information Technology*, Vol. 28, No. 5, (Sept. 2009), pp. 405-419, ISSN 0144-929X.

Toyama, K.; Krumm, J.; Brumitt, B. and Meyers, B. (1999). Wallflower: Principles and practice of background maintenance. *The Proceedings of the Seventh IEEE International Conference on Computer Vision*, pp. 255-261, ISBN 0-7695-0164-8, Kerkyra, Greece, Sept. 1999, IEEE Computer Society, CA, USA.

Vincent, L. (1993). Morphological grayscale reconstruction in image analysis: applications and efficient algorithms. *IEEE Transactions on Image Processing*, Vol. 2, Iss. 2, (April 1993), pp. 176-201, ISSN 1057-7149.

White, B. & Shah, M. (2007). Automatically tuning background subtraction parameters using particle swarm optimization. *In Proc. of the IEEE Int. Conference on Multimedia and Expo*, pp. 1826-1829, ISBN 1-4244-1016-9, Beijing, China, July 2007, IEEE Computer Society, CA, USA.

Wren, C.R.; Azarbayejani, A.; Darrell, T.J. & Pentland, A.P. (1997). Pfinder: Real-time tracking of the human body. *IEEE Transactions on Pattern Analysis and Machine Intelligence*, Vol. 19, No. 7, (July 1997), pp. 780-785, ISSN 0162-8828.

Zhong, J. & Sclaroff, S. (2003). Segmenting foreground objects from a dynamic, textured background via a robust Kalman filter. *The Proceedings of the Ninth IEEE International Conference on Computer Vision*, pp. 44-50, ISBN 0-7695-1950-4, Nice, France, Oct. 2003, IEEE Computer Society, CA, USA.

Zhou, Q. & Aggarwal, J. (2001). Tracking and classifying moving objects from videos. In *Proc. of the 2001 IEEE Workshop on Performance Evaluation of Tracking and Surveillance*, pp. 1-7, Kauai, Hawaii, USA, Dec. 2001.

Zhou, J.; Gao, D.; & Zhang, D. (2007). Moving Vehicle Detection for Automatic Traffic Monitoring. *IEEE Transactions on Vehicular Technology*, Vol. 56, No. 1, (Jan. 2007), pp. 51-59, ISSN 0018-9545.

Zhu, Z. & Xu, G. (2000). VISATRAM: A real-time vision system for automatic traffic monitoring. *Image Vision Computing*, Vol. 18, No. 10, (July 2000), pp. 781–794, ISSN 0262-8856.

# 5

# Urban Air Quality and Road Traffic Air Pollution Modelling of Szeged

Tamás Weidinger[1], Györgyi Baranka[2],
László Makra[3] and András Zénó Gyöngyösi[1]
[1]*Eötvös Loránd University, Department of Meteorology*
[2]*Hungarian Meteorological Service*
[3]*Department of Climatology and Landscape Ecology, University of Szeged*
*Hungary*

## 1. Introduction

Monitoring the actual state and tendencies of urban air quality and its forecasting plays an important role in the environmental and meteorological research of Hungary, since it is a key element of the urban life quality and health protection. After the change of the political system in 1990, with the modernization of the industry, road traffic became the most important pollution source: for example, the contribution of road traffic to the total air pollution in large and medium cities such as Budapest and Szeged is between 70 and 80%.

Similarly to the common prediction of environmental and meteorological parameters, their measurements also constitute a closed system (Baklanov, 2006; Baklanov et al., 2006; Bozó et al., 2006a,b). The importance of regional and urban scale studies, environmental plans, impact assessments, statistical and dynamical modelling of air pollution and prediction of hazardous meteorological situations have grown significantly beside large- and meso-scale air quality modelling.

The University of Szeged established its importance in the research of the urban heat island and the discovery of relations between meteorological parameters and air pollution (Makra et al., 2006a; 2006b; Unger & Pongrácz, 2008). In the scale dependent modelling of dispersion the Hungarian Meteorological Service, the Department of Meteorology at Eötvös Loránd University and the Department of Fluid Mechanics at Budapest University of Technology and Economics are the most important institutes (Steib & Labancz, 2005; Lovas et al., 2006; Steib et al., 2008; Balczó et al., 2009).

Following the review of air quality research in Hungary, we introduce the national pollution monitoring network; afterwards we deal with the opportunity of urban air quality modelling. The correlation between urban air quality and the large-scale weather situations (macrocirculation patterns) in long term trends of the most important pollutants and road traffic emissions will be demonstrated through the example of Szeged.

The city of Szeged (20°06' E; 46°15' N) is the largest town in southeast Hungary, and located at the confluence of the rivers Tisza and Maros. The built-up area covers a region of about 46 $km^2$ with 155,000 inhabitants. The major part of air pollution occurs in the urban area as a consequence of road traffic and heating (ca. 70% of dust is emitted on the roads). The amount of pollutants emitted by industry or advected by continental or regional transport has been less significant. Traffic data includes traffic census, speed and vehicular emission data, while the meteorological data consists of (i) daily macrocirculation patterns, (ii) synoptic, road and urban meteorology data. The effects of traffic emissions on the ambient air quality in Szeged were evaluated using the CALINE4 and the CAR urban air pollution models. Both models are Gaussian type dispersion models specifically designed for the evaluation of air quality impacts of roadway projects. Each roadway link is treated as a separate emission source, producing a plume of pollutants which disperses downwind. Air pollution concentrations at any specific location are calculated using the total contribution from the overlapping pollution plumes originating from the sequence of roadway links. Both models had been run on realistic input parameters, regional and city background concentrations. Street geometry and traffic data for the period 1997–2008 at Szeged have been used. The annual mean concentrations on some major roads of the Szeged region clearly indicate the increase of CO levels and a much more definite increase of $NO_2$ concentrations between the beginning and end of the period examined. Model results have been compared to measurements showing good agreement with (i) slight overestimation of concentration due to the insufficient consideration of technological development of vehicles using the CAR model; (ii) remarkably higher vehicle-originated air pollution, as clearly indicated by the CALINE4 dispersion model.

## 2. Urban and rural air quality research: a historical overview

The atmospheric chemistry research and studies of urban air quality have a history of about one and a half century. The Institute of Public Health in Budapest was established by József Fodor in 1874. Similarly to the program of the French observatory in Montsouris, besides dust, measurement has been done for the assessment of carbon-dioxide, ozone and ammonia.

Till the end of the 1930s, due to industrial activity and heating mostly based on burning brown coal with a high sulphur content, sulphur-dioxide became the most important pollutant, with mean annual concentrations higher than 2000 $\mu g\ m^{-3}$ in the most polluted parts of the capital which decreased to 900 $\mu g\ m^{-3}$ for industrial regions and 240 $\mu g\ m^{-3}$ for the whole area of Budapest in the first part of the 1960s (Mórik, 1970). The annual concentrations of $SO_2$ for Budapest and Szeged were 4.9 $\mu g\ m^{-3}$ and 6.3 $\mu g\ m^{-3}$, respectively, in 2009. This was the result of changing the system of heating and the structure of industry.

The urban air quality monitoring system had been established by the end of the 1970s. Nowadays, pollutants of road traffic (e.g.: $NO_x$, CO) have become dominant, but the dust and soot for Budapest and the dust for Szeged are also significant loads.

The immission thresholds for the most important pollutants are presented in Table 1. The fact that Szeged has been pushed into the list of clean air cities, compared to other Hungarian towns with heavy industries in the second half of the 20$^{th}$ century, shows the change in the structure of industry and road traffic. Time variation of standard urban pollutant concentrations from 2002 based on measurements of the Hungarian Air Quality Network using data from all stations are shown in Table 2. No significant trend can be detected in the trace gas concentrations. The amounts of $PM_{10}$ and $PM_{2.5}$ are slightly decreased.

Background air quality measurement and modelling are also the scope of meteorology. One of the most important trace gases is ozone. Historical ozone data sets are available from the mid 1800s for Buda and Szeged. The first attempts in atmospheric chemistry and atmospheric physics were made at the Main Meteorological and Geomagnetic Observatory

established in 1867 at Ó-Gyalla (today Hurbanovo, Slovakia). The almanacs of the observatory contained ozone measurements beside meteorological and geomagnetic data between 1893 and 1905 for day- and night-time on the 10-degree Schönbein scale.
The chemical measurements of precipitation were initiated at the beginning of the 20th century. Comparing to data from the 1980s, Horváth (1983) pointed out that the anthropogenic emission of $NO_x$, (burning of fossil fuels, emission from vehicles) had been markedly increasing since the beginning of the 20th century. The nitrate concentration measured in 1902 (0.13 mg N $litre^{-1}$ in the precipitation) was only 15% of the level of the early 1980s (0.86 mg N $litre^{-1}$). This value has been measured to be about 2.42 mg N $litre^{-1}$ (at station K-puszta, Fig. 1) in 2009, a large increase.

| Air pollution [$\mu g\ m^{-3}$] | Highly Protected | | | Protected I. | | | Protected II. | | |
|---|---|---|---|---|---|---|---|---|---|
| | **year** | **24 h** | **0.5 h** | **Year** | **24 h** | **0.5 h** | **Year** | **24 h** | **0.5 h** |
| $SO_2$ | 30 | 100 | 150 | 70 | 150 | 500 | 150 | 500 | 1000 |
| CO | 1000 | 2000 | 5000 | 2000 | 5000 | $10x10^3$ | 5000 | $10x10^3$ | $20x10^3$ |
| $NO_x$ | 30 | 70 | 85 | 100 | 150 | 200 | 150 | 200 | 400 |
| $NO_2$ | 30 | 70 | 85 | 70 | 85 | 100 | 120 | 150 | 200 |
| Dust | 30 | 60 | 100 | 50 | 100 | 200 | 100 | 200 | 300 |
| Dust deposition | 30 days: 12 g $m^{-2}$<br>year: 100 t $km^{-2}$ | | | 30 days: 16 g $m^{-2}$<br>Year: 120 t $km^{-2}$ | | | 30 days: 21 g $m^{-2}$<br>year : 150 t $km^{-2}$ | | |

Table 1. Immission concentration limits in Hungary

Regular precipitation chemistry measurements were only initiated in 1965. The background concentrations of sulphur-dioxide and nitrogen-oxides have been measured since 1973. Currently, several background atmospheric precipitation chemistry and/or atmospheric pollution stations (Fig. 1) have been operating in Hungary (Haszpra et al, 2006; Weidinger et al., 2010). Only one of them, located at K-puszta, has an overall importance: it has been sending precipitation chemistry, trace gas, aerosol and heavy metal concentration measurement data to the EMEP (European Monitoring and Evaluation Programme) centre since its establishment in 1974 (Müller et al., 2007; Haszpra, 2008; Strategy for EMEP 2000–2009). This station also yields background concentration data to the air quality measurements in Szeged.

| $\mu g\ m^{-3}$ | **$SO_2$** | **$NO_2$** | **CO** | **$O_3$** | **$PM_{10}$** | **$PM_{2.5}$** |
|---|---|---|---|---|---|---|
| 2002 | 5.5 | 35 | 551 | 24 | 28.4 | |
| 2003 | 10.3 | 46 | 646 | 44 | 27.6 | |
| 2004 | 7.8 | 43 | 565 | 43 | 25.3 | |
| 2005 | 8.1 | 44 | 859 | 43 | 29.3 | |
| 2006 | 8.8 | 42 | 776 | 44 | 29.3 | 21.1 |
| 2007 | 7.4 | 37 | 608 | 50 | 26.8 | 17.6 |
| 2008 | 7.2 | 35 | 452 | 51 | 20.3 | 14.2 |
| 2009 | 6.1 | 34 | 433 | 46 | 25.4 | 16.6 |

Table 2. Time variation of standard urban pollutant concentrations from 2002 based on measurements of the Hungarian Air Quality Network, annual averages include measurement results of all sites of the network, calculated by Lautner (2009).

Regulatory air pollution modelling has been carried out in Hungary since the early 1960s. Initially Gaussian puff models were used, in which stability categories and boundary layer depths were calculated. For the uniform application of transmission schemes the standardization of air quality models is crucial. This work ended in the early 1980s, when pollution of point, line and aerial sources were modelled. These models can be applied even in urban environments (Fekete et al., 1983; Gyöngyösi et al., 2009). The next step was the development of the Hungarian Standardized Model in the 1990s (HNS-TRANSMISSION). This Gaussian transmission model can consider contribution of up to 50 sources. It is suitable to describe transmission processes from local to regional scales including the effect of orography (Szepesi et al., 2005).

The EPA AERMOD system has been implemented at the Hungarian Meteorological Service as a powerful tool for case study calculations (Steib, 2005). The development of a meteorological pre-processor for the model has also been performed. Several dispersion models have been applied in the urban environment in Hungary, such as ADAMS, AURORA, CAR, CALINE4, ISC3, MISCAM, among others, but even wind tunnel experiments have been performed and the FLUENT numerical solver has also been applied for urban air pollution research purposes. Several models of industrial accident risk assessment have been run. Input data for these models are yielded by the prediction system of the Hungarian Meteorological Service from the ALADIN and MM5 model outputs and the forecast datasets of the ECMWF centre (Hanna et al., 2001; Havasi & Zlatev, 2002; Moussiopoulos et al., 2003; Mensink et al., 2008; Kristóf et al., 2009; Mészáros et al., 2010). WRF, earlier the ETA numerical weather prediction model systems are also used for air pollution modelling purposes (Weidinger et al., 2006).

Measurements of rural and urban air quality constitute a uniform system with meteorological information. The Hungarian Meteorological Service provides a complex database for rural and urban environmental and standard meteorological measurements. This job is performed by the Air Quality Reference Centre, which was established in 2009.

## 3. The air pollution monitoring system in Hungary

Currently, the existing network in Hungary comprises 37 fully automatic monitoring stations for which the Hungarian Meteorological Service (http://www.kvvm.hu/olm/) is responsible (data has been available since 1974). Station locations are shown in Fig. 1. Stations monitor air pollutants as well as meteorological parameters and are connected to a central automatic collection and processing unit. The monitoring network is operated, managed and maintained exclusively by the Hungarian Meteorological Service. All data collection, processing and maintenance operations are performed by the Hungarian Air Quality Network's (HAQN) officials and technical personnel. They are responsible for issuing short term emergency bulletins on days of high pollution levels and imposing various restrictive measures to limit the magnitude of pollution episodes. Long-term strategic policy planning and implementation of permanent measures are also the responsibility of HAQN. The assessment of air quality is based upon monitoring and modelling at different temporal and spatial scales for regulated components in Europe: $PM_{10}$ (and $PM_{2.5}$), NO, $NO_X$, $NO_2$, CO, $SO_2$, $O_3$ and benzene. With manual sampling method at 199 measuring points $SO_2$ and $NO_2$ measurements are carried out daily and collected weekly. At 296 measuring points, dust deposition is detected on a monthly basis.

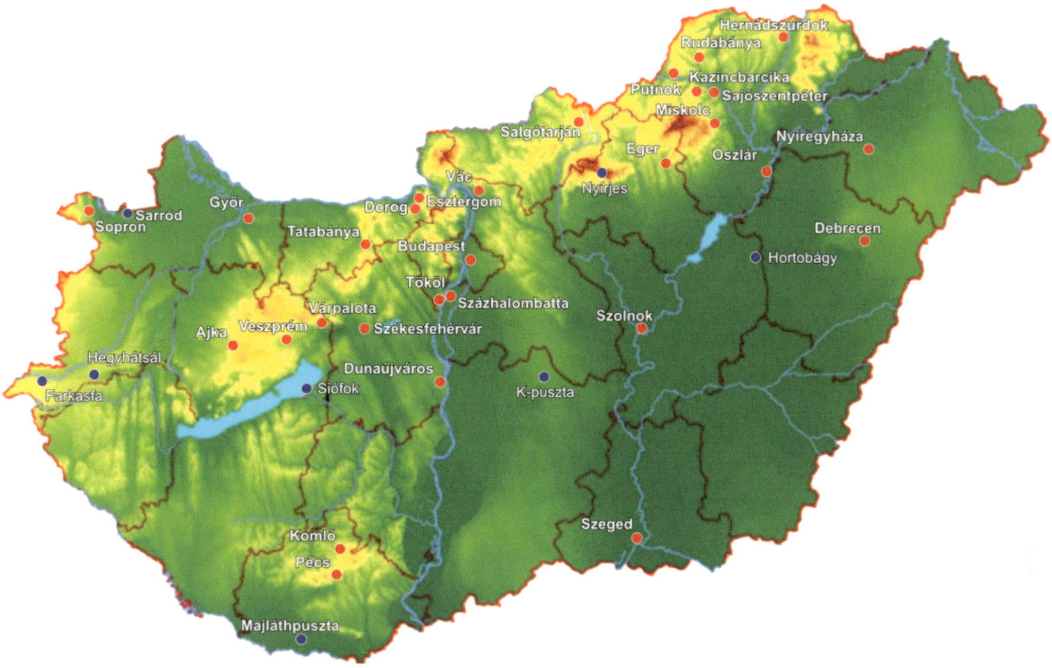


Fig. 1. Monitoring network of regional background air pollution (blue dots) and urban monitoring sites (red dots) in Hungary

## 4. The variation of air pollution in terms of ozone concentration in the past century and a half

Ozone, one of the most important trace gases in the atmosphere, was discovered by Christian Friedrich Schönbein (1799–1886) Professor of Chemistry at the University of Basel. It is a secondary pollutant, which means that it does not have direct sources near the surface. The level of ozone concentration was about 5 to 15 ppb (10 to 30 $\mu g\ m^{-3}$) in the second half of the 19th century with a weak daily and annual variation. As a result of anthropogenic effects (caused mainly by road traffic), concentration values have since increased to double or triple that. Ozone is harmful to living tissues even at low concentrations. Humans have become adapted to this load (Möller, 1999), but its very high current concentration can no longer be tolerated.

Ozone is generated by photochemical reactions of anthropogenic pollutants (eg. NO, $NO_2$, CO, VOC) and natural organic compounds (isoprene). Another important source for tropospheric ozone is the down-mixing of stratospheric ozone during tropopause folding events. The near surface ozone concentration has a typical annual and daily variation. Its annual variation can be characterized by a summer maximum and winter minimum. A secondary maximum can occur during early spring due to the down-mixing of stratospheric ozone. The highest near surface concentrations may occur in the afternoon and early night hours.

Let us now consider historical measurements (Fig. 2). Measurements of atmospheric ozone were made using Schönbein's method at more than 20 stations in the Habsburg Empire in the period 1853–1856, which resulted in the most detailed database ever made, in which

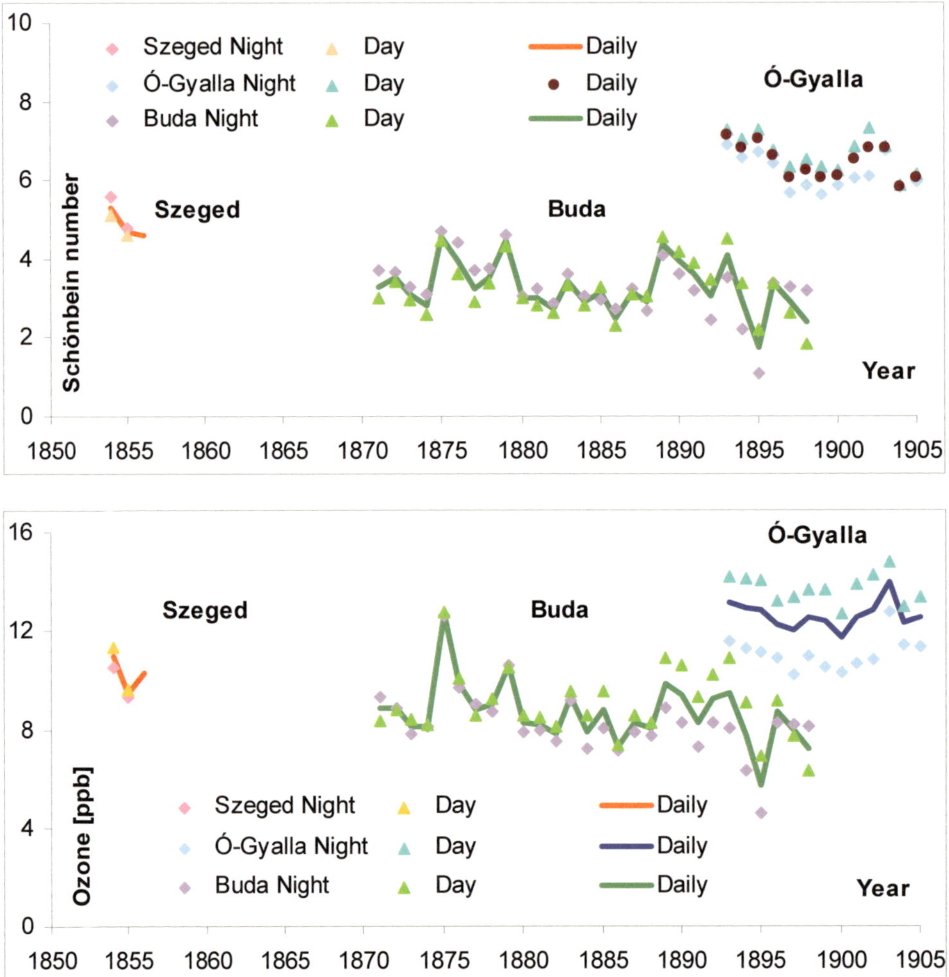


Fig. 2. Annual mean Schönbein numbers (top) and ozone concentrations (1 ppb ~ 2 $\mu g\ m^{-3}$) at Szeged, Ó-Gyalla (presently in Slovakia: Hurbanovo) and Buda (bottom), respectively, for the period 1854–1905 (day-time, night-time and daily mean values)

meteorological data are presented together with Schönbein numbers. In the territory of the Hungarian Kingdom ozone measurements were performed, *inter alia*, in Buda (its German name in the database: Ofen), Szeged (Szegedin), Selmecbánya (Schemnitz) and Besztercebánya (Neusohl). Long-term datasets from Buda (1871–1898) and Ó-Gyalla (1898–1905) are also available in the archives of the Hungarian Meteorological Service and in the library of the Faculty of Science, Eötvös Loránd University (Weidinger et al., 2009). Ozone concentration has been determined following Pavelin et al., (1999), based on day-time and night-time Schönbein numbers and relative humidity data.

Our results for Szeged, Buda and Ó-Gyalla, including annual average Schönbein numbers and ozone concentrations between the middle of the 19th century and the beginning of the 20th century are presented in Fig. 2. The highest Schönbein number values and ozone concentrations were obtained for Ó-Gyalla and the lowest ones for Buda. There is a slight

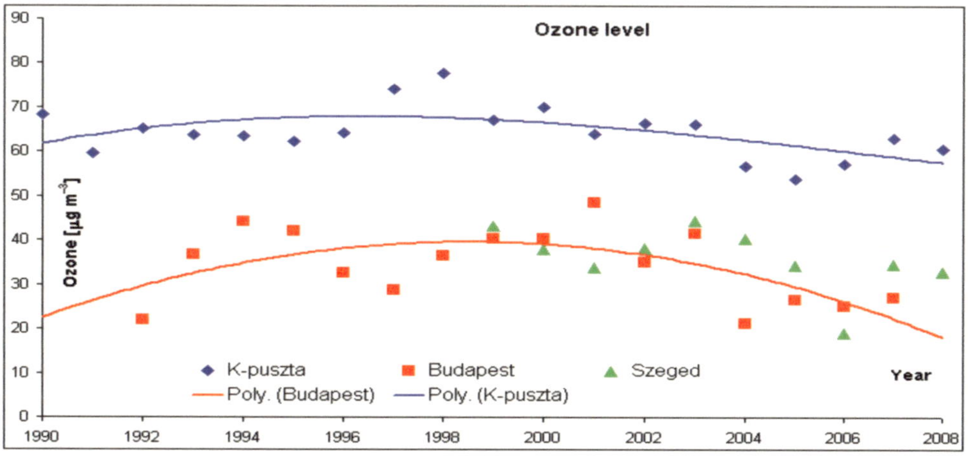


Fig. 3. Annual mean ozone concentration in Budapest and Szeged (both urban) and K-puszta (rural) stations, 1990–2008

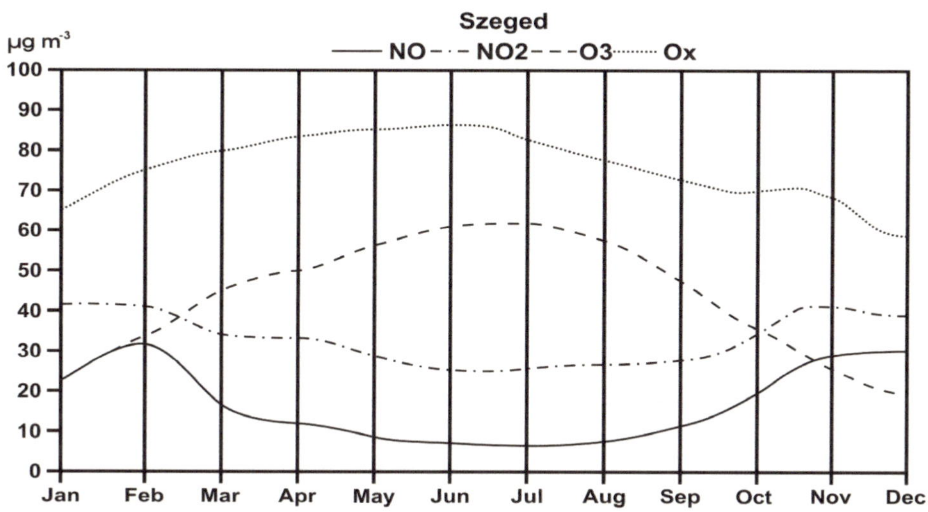


Fig. 4. Time variation of the annual mean concentration of NO, $NO_x$, $O_3$ and Ox, respectively, at Szeged between 1997 and 2001 (Kiss et al., 2005)

difference between the values measured at Buda and those measured at Szeged. No significant trend can be found in the data series at Buda, but the population and correspondingly economic potential of Budapest (after incorporating cities Óbuda, Buda and Pest in 1873, the name of the capital became Budapest) grew dramatically. Simultaneous measurements at Buda and Ó-Gyalla represent the difference between rural and urban environmental conditions. The course of annual means indicates an inter-annual variation of about 8–12 ppb (16–24 μg $m^{-3}$) at Szeged and Buda, and about 12–14 ppb (24–28 μg $m^{-3}$) at Ó-Gyalla. Ozone concentrations have been measured to be higher in the day-time and lower at night. The difference between day-time and night-time data is lowest in Szeged, and the highest (especially at the end of the considered period) in Buda and Ó-Gyalla, which can be explained by the effect of urbanization.

It is also obvious that values measured in Ó-Gyalla are higher than those in Buda within the same period. This highlights a phenomenon that has been experienced even today: in the clean air of remote sites, far from industrial pollution sources, higher ozone concentrations can be recorded due to the different vegetation. After performing a more detailed analysis of the data series from Buda, the day-time concentrations in spring and summer turned out to be usually higher than those at night, which shows the effect of photochemical processes. In autumn and winter, however, days with higher nocturnal concentration occur more frequently. A possible explanation is lower insolation, the effect of vegetation and coal heating.

Comparing historical data to current measurements, the trend of the difference in the urban-rural ozone concentration is becoming larger and larger; furthermore, an ever increasing difference can be observed between day-time and night-time concentrations as well. The role of photochemical processes has become more pronounced with the increase of the emission of nitrogen-oxides and other trace gases that are important in ozone generation (Bozó et al., 2006b). Daily variations are also increasing.

The annual mean concentrations between 1990 and 2008 at Budapest, Szeged and K-puszta are shown in Fig. 3. Note that until 1998 the tendency was positive, while since 1999 concentrations have been decreasing at all three stations. The data of Szeged fit the values of Budapest (urban stations). Current measurements indicate almost three times higher ozone concentrations compared to those in the $19^{th}$ century.

Anthropogenic effects and the increased photochemical processes exert substantial influence on the monthly values of $O_3$, NO, $NO_x$, Ox, respectively (Fig. 4). The primary reason for the growing ozone and Ox concentrations is the increasing density of road traffic (Makra et al., 2003; Kiss et al., 2005). Ox is a measure of oxidant concentrations, represented by the sum of $O_3$ and $NO_2$ concentration contained in an air mass (after Warnek, 1999). It is more suitable for the assessment of the photochemical $O_3$ budget than $O_3$ alone, because it takes ozone-related reversible chemical processes into account as well (Brönnimann & Neu, 1997; Mayer, 1999). The annual course of ozone concentration is connected with the photochemical processes. These, along with road traffic emissions and turbulent mixing, explain the annual course of NO and $NO_x$ concentrations (maxima in winter, minima in summer). The shape of the annual course and the values of the trace gases examined (Fig. 4) have not changed during the last few years (see also Fig. 3 and Table 2).

## 5. Dynamic and statistical modelling of urban air quality

Air pollution levels depend on (i) total emissions, (ii) transport, (iii) chemical transformation phenomena, and (iv) deposition processes in the atmosphere. The complexity of the overall issue calls for the use of mathematical tools, the so-called air pollution models. There are a wide variety of model types available for use in air quality modelling.

Air quality models can be classified with respect to the scale of the phenomena (see Fig. 5) they are developed to simulate. For urban areas we distinguish local scale models - which include street canyon models - from urban scale models. The model domain of a street canyon model is the size of a street. The obstacles (i.e., buildings) are resolved. The model domain of a local scale model is a town or city district (up to 10 km x 10 km). The phenomena that can be resolved by local scale models have to cover a lifetime of at least several minutes and a horizontal scale of several metres. Integration time for one-day simulation should be performed in a few hours.

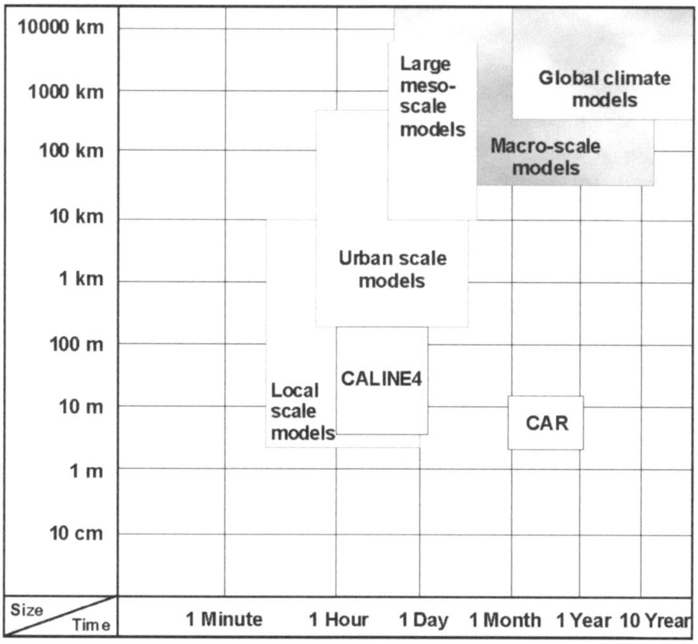


Fig. 5. Spatial and temporal scales of air quality models based on Moussiopoulos et al. (2003). Resolution of CALINE4 and CAR traffic models is also indicated

Urban scale and other larger scale models do not explicitly treat each building, but include the effect of them by using a corresponding roughness length. The model domain of an urban scale model is the size of an EU country or part of one (up to 500 km x 500 km). Urban scale models may be integrated for several days when using a nesting approach. They can resolve phenomena that have a characteristic time scale of more than 10 minutes and a characteristic horizontal scale of more than several hundreds of metres. The synoptic scale model domain covers the European-Atlantic region (5000 km x 5000 km) and the macro-scale models have a model domain that covers part of the globe (larger than 5000 x 5000 $km^2$).

Models can be generally divided into four categories, based on the methods used for the calculations: 1) statistical analysis models, 2) Gaussian models and 3) Lagrangian models. Most models, however, use an 4) Eulerian approach to calculate pollution transport. Some models may make use of more than one methodology and may fall into more than one category.

*1 Statistical*

Statistical models are based on relationships, usually originally founded in physical concepts, between observations and some other relevant parameters. There are a wide variety of statistical methods available including regression, neural networks, etc. that can be used. Since the relationships are based on observations, statistical models need to be established in each environment in which they are used. Statistical models simply require single-valued meteorological data extracted from a coupled meteorological model. They are not usually suitable for carrying out mitigation planning, as they do not causally relate emissions to concentrations.

*2 Gaussian*

These types of models solve analytical equations that are usually approximations to the physical processes governing the chemistry and/or transport and diffusion. A typical example of this type is the Gaussian plume model, which assumes that turbulent dispersion can be described using a Gaussian distribution profile. These types of models are often used since they have been found under many circumstances to provide good estimates of concentrations from industrial or traffic sources. Other types of models that fall into this category are well-mixed box models or other non-Gaussian types that include higher order moments in their dispersion description. There are also a number of analytical models that may provide statistical assessments. Gaussian models used for screening applications for industries are often of this type. These models require meteorological data for a single point or possibly a vertical profile and the evaluation of the Monin-Obukhov turbulence scaling parameters. Even if the interface module computations required by the previous classes are quite limited, the extraction of three-dimensional meteorological data representative of conditions assumed to be uniform over the whole urban area is quite critical, especially for large cities located in complex terrain.

*3 Lagrangian*

Lagrangian models can refer to a wide variety of models but they all encapsulate the concept that the model describes, or utilizes, trajectories of single or multiple particles as the reference frame for the model. These models need: 3D fields of average quantities like wind, temperature, humidity and possibly turbulent kinetic energy; 2D surface fields like precipitation, sensible heat flux, friction velocity and Monin-Obukhov length; 3D turbulence fields, like wind variances and Lagrangian time scales, that have to be evaluated from mean variables or reconstructed from boundary layer scaling parameters.

The chemistry models applied range from no chemistry via simple gas-phase chemistry considering NO, $NO_2$ and $O_3$, to complex gas phase reaction schemes. Aerosols are mostly treated as being passive but some steps are taken to consider chemically active aerosol or aqueous phase chemistry. Most models use a resistance model for calculating dry deposition, but only very few models consider wet deposition for different pollutants.

*4 Eulerian*

Eulerian models are the most widely used models for urban and regional scale applications. Equations of motion, chemistry and other physical processes are solved at regular fixed points in space. The Eulerian grid can be, as for regional scale models, based on geographic coordinates such as latitude and longitude, or on local coordinates for distance. For urban scale applications, grid sizes are generally in the range of 1–5 km and for regional scale applications the range is 10–50 km. High resolution Eulerian models are also available, which can solve the equations of motion down to metre-scale, or even less. These types of models, Large Eddy Simulation (LES) or Computational Fluid Dynamics (CFD) models, can resolve buildings and other obstacles and allow a level of detail at hot-spots that cannot be achieved using the larger meso-scale models. 3D Eulerian models need the 3D average meteorological fields. The Eulerian horizontal and vertical dispersion coefficients ($K_H$, $K_Z$) produced by numerical weather prediction models can be directly used in these models. Nevertheless this practice is not always possible or advisable, and therefore the interfaces for Eulerian models are usually implemented with capabilities to re-compute turbulence parameters from mean meteorological variables and scaling parameters.

Though the European Air Quality Directives do not specify any selection criteria among air pollution transport models other than the quality objectives, it is understood that some models are more suitable than others for the assessment applications. All types of models are presently used for air quality management and forecasting in urban areas at European cities. The latest-generation steady-state models (Helsinki and Bologna) and Eulerian Chemical Transport Models (Budapest, Oslo, Turin, London and Castellon/Valencia) have been implemented into the respective air quality forecasting systems, while Lagrangian models are used for emergency preparedness systems (Copenhagen).

The regulatory line source models can be broadly classified as Gaussian, Eulerian or Lagrangian models. Lagrangian dispersion models (like GRAL) have become increasingly more feasible, due to the advances in computer technology (Oettl et al., 2001a). Computations using Lagrangian dispersion models are still limited by computing capacity usually available for regulatory dispersion studies. The ROADWAY (Eskridge & Catalano, 1987) and MGO (Berlyand et al., 1990) models are based on a K-theory (Eulerian) approach. An obvious advantage of the K-theory models is that they can readily include the interaction of diffusion processes and chemical transformation. Both Eulerian and Lagrangian models are less limited by topographical and meteorological conditions than Gaussian plume models (Oettl et al., 2001b). Different versions of the Gaussian line source model have been used for dispersion evaluations from a road. Such models include HIWAY-2 (Petersen, 1980), CALINE4 (Benson, 1986 and 1992), GM (Chock, 1978), GFLSM (Luhar & Patil, 1989) and OMG (Kono & Ito, 1990).

## 6. Urban climate and urban air quality research in Szeged

### 6.1 Geographical location and air quality conditions of Szeged

The city of Szeged, the largest town in southeast Hungary, (20°06' E; 46°15' N) is located at the confluence of the rivers Tisza and Maros, characterized by a landscape of extensive flats and an elevation of 79 m a.s.l. (Figs. 6–7). Szeged and its surroundings are not only characterized by extensive lowlands but the city also has the lowest elevation of any city in the Carpathian Basin, putting it in a special, so-called "basin-in-abasin" or "double basin" situation. This special situation is also favourable to the development of stronger anticyclonic situations, enabling the enrichment of pollutants in the air.

In a detailed analysis, Szeged was ranked 32$^{nd}$ out of 88 Hungarian cities by the quality of its environment and its level of environmental awareness. [The city ranked 1$^{st}$ was considered to have the best environmental conditions (Makra et al., 2002).]

The air quality of Szeged is basically determined by nitrogen oxides ($NO_x$) (Fig. 4), ozone ($O_3$) (Fig. 4) and particulate matter ($PM_{10}$) (Fig. 8). The concentrations of these parameters tend to exceed the air quality limit values of EU standards (Fig. 4 and Fig. 8). [The daily (24-hour) concentration of particulate matter in few cases is 11–19 times higher, while its annual concentration is twice the EU standard (highly protected area).] The high concentrations of particulate matter are closely correlated to the development of respiratory diseases. The annual trends of the air pollutant levels follow a unimodal distribution. The concentrations of NO, $NO_2$ and $PM_{10}$ are characterized by summer maxima and winter minima. At the same time, ozone reaches its maximum in the summer, in accordance with the annual change of irradiation (Makra & Horváth, 2001; Makra et al., 2001a; 2001b; Mohl et al., 2002; Mayer et al., 2004).

About 50% of the particulate matter comes from the region northwest of Szeged covered by shifting sands, loess and sandy ridges. At the same time, the industrial district is found at

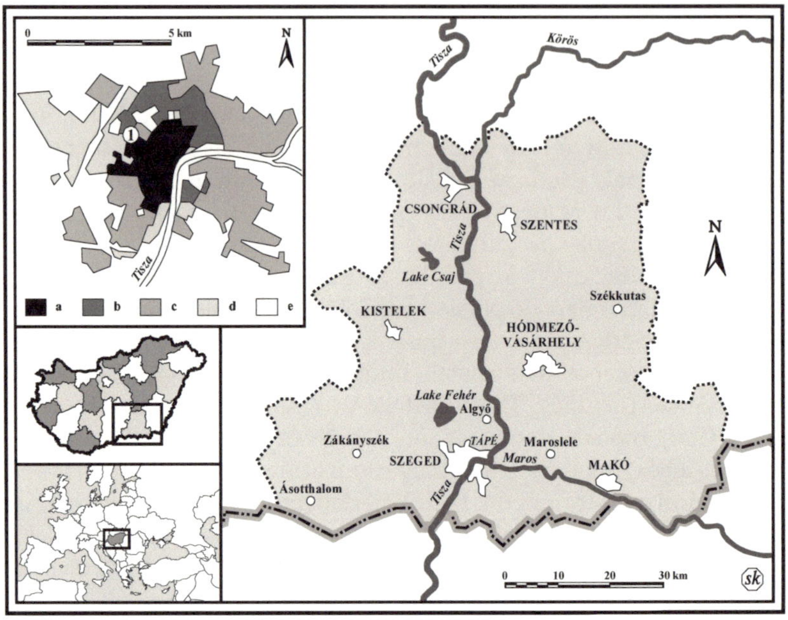


Fig. 6. The geographical position of Szeged, Hungary and types of built-up areas of the city (top left) [a: city centre (2-4-storey buildings); b: housing estates with prefabricated concrete slabs (5-10-storey buildings); c: detached houses (1-2-storey buildings); d: industrial areas; e: green areas; (1): monitoring station]

Fig. 7. Instruments of the monitoring station. Measurement program: $O_3$, NO, $NO_2$, $SO_2$, CO, $PM_{10}$ and standard meteorological measurements (temperature, relative humidity, wind speed and direction, global radiation and precipitation)

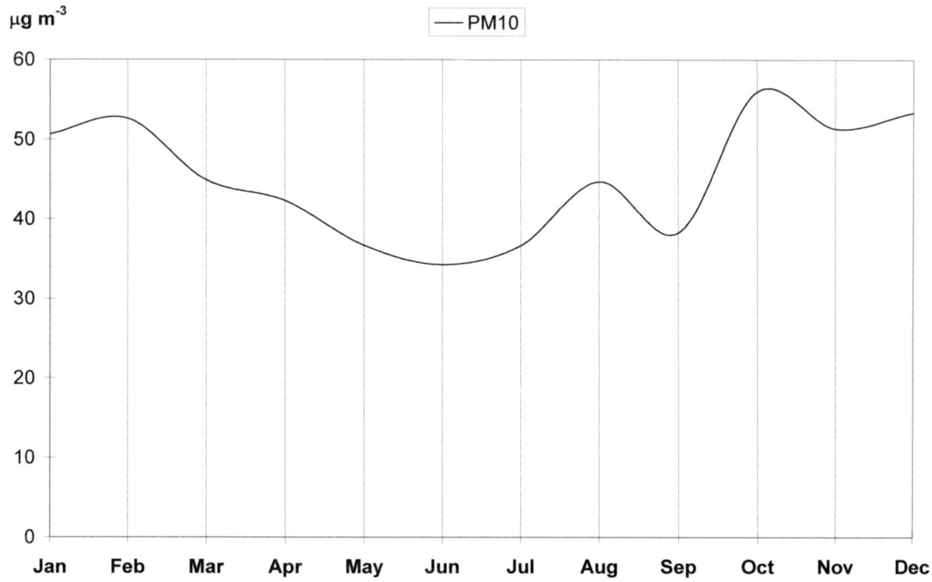


Fig. 8. Average annual variations of $PM_{10}$, monitoring station, Szeged, 1997-2001

the northwest part of the town. Hence, the dominant north-westerlies transport particulate matter as well as pollutants of industrial origin over Szeged. The rest of $PM_{10}$ originates in traffic. Particulates are produced partly by vehicle engines and partly by air currents when passing vehicles churn dust into the air (Mohl et al., 2002).

Another serious environmental health problem arises during late summer and early autumn caused by the pollen of ragweed with mugwort leaves or short ragweed (*Ambrosia artemisiifolia* = *Ambrosia elatior*), considered the most dangerous of all pollens. The annual pollen counts of ragweed in Szeged are one or two orders of magnitude higher than those in other European cities. Hence, in this respect, Szeged is considered one of the most polluted cities in the continent (Makra et al., 2004; 2005).

The objectives of this section are: (i) to determine the typical diurnal, weekly and annual variations of traffic-related air pollutants NO, $NO_2$, $O_3$, Ox and $PM_{10}$ in Szeged, (ii) to determine statistical interrelationships of these air pollutants, furthermore, (iii) as a case study, to analyse their connection to meteorological elements for Szeged, by using the dataset of self-performed traffic census. On the basis of these datasets we analyse the interrelationship of traffic density, pollutant levels and meteorological elements. Further points of the analysis are evaluating the air quality of Szeged using different indices, comparison of pollutant characteristics on weekdays vs. non-weekdays, expressing the efficacy of classifying air pollutants on objective and subjective weather types, expressing the role of the meteorological elements in enriching/diluting pollutant concentrations and modelling the air quality of Szeged by applying CALINE4 and CAR models and comparing their results.

### 6.2 Location of the monitoring station and the database

The monitoring station is located in downtown Szeged in a crossroad with heavy traffic (Kossuth Avenue and Damjanich Street), about 15 m distance from Kossuth Avenue and ca. 30 m away from the road axis. This is one of the busiest crossroads of Szeged. The monitoring station was put into operation on September 1, 1996. Sensors, measuring

concentrations of air pollutants, are placed 3 m above the surface. Values of the meteorological elements and concentrations of the main air pollutants (CO, NO, $NO_2$, $SO_2$, $O_3$ and $PM_{10}$) are measured at the monitoring station (Fig. 7).
The following meteorological and pollution parameters are used in the study. The 12 meteorological parameters used are as follows: mean temperature ($T_{mean}$, °C), maximum temperature ($T_{max}$, °C), minimum temperature ($T_{min}$, °C), daily temperature range ($\Delta T = T_{max} - T_{min}$, °C), wind speed (WS, m $s^{-1}$), relative humidity (RH, %), irradiance (I, MJ $m^{-2}$ $day^{-1}$), saturation vapour pressure (E, mm Hg), water vapour pressure (VP, mm Hg), potential evaporation (PE, mm), dew point temperature ($T_d$, °C) and atmospheric pressure (P, mm Hg). The 8 pollution parameters considered are the average diurnal mass concentrations of the following pollutants: CO (μg $m^{-3}$); NO (μg $m^{-3}$), $NO_2$ (μg $m^{-3}$), $SO_2$ (μg $m^{-3}$), $O_3$ (μg $m^{-3}$) and $PM_{10}$ (μg $m^{-3}$) as well as the daily ratios of $NO_2$/NO and the daily maximum concentrations of $O_3$ (μg $m^{-3}$).

### 6.3 Results

#### Comparison of the annual variation of the pollutant concentrations and its effect on variability

The diurnal concentrations of NO display a clear annual variation, while those of $NO_2$ and $PM_{10}$ indicate less definite variation with much higher fluctuation. However, annual variations for all three are characterized by winter maxima and summer minima. In contrast, the diurnal concentrations of $O_3$ with a clear annual variation have a winter minimum and a summer maximum.
The average weekly variations of the air pollutants, after the annual variation had been removed, were also determined. The highest concentrations are observed in winter and the lowest ones in summer. On the other hand, the varitation of $O_3$ concentrations has weekday minima and weekend maxima. On weekdays the concentration of traffic-related NO is high. After reacting with $O_3$, its concentration decreases:

$$NO+O_3 \longrightarrow NO_2+O_2 .$$

Conversely, at the weekend, the concentration of $O_3$ is high, due to the relatively low traffic. Fig. 4 represents the average annual variations of NO, $NO_2$, $O_3$ and Ox. The annual variation of the primary air pollutant NO displays the greatest values in November, December and January, with a maximum in February (Fig. 4). As the NO concentration depends not only on the rate of emission, but on the prevailing weather conditions as well, higher winter values refer to atmospheric stability with frequent inversions. The average annual variation of NO has the lowest values during the summer (June and July), when air mass exchange is the most intensive. The annual variation of $NO_2$, a secondary substance produced mainly by chemical reactions, follows a similar course to that of NO (Fig. 4). Tropospheric ozone is produced via the effect of short-wave radiation on substances emitted from anthropogenic sources. The role of solar radiation in the troposphere, producing photochemical $O_3$, can be expressed by the following pair of chemical equations:

$$NO_2 + h\nu \longrightarrow NO + O ,$$

$$O + O_2 + M \longrightarrow O_3 + M ,$$

(*h*: Planck-constant; *ν*: frequency of irradiance; *M*: usually a molecule of $O_2$ or $N_2$) (Sindosi et al, 2003). [Simple chemical reactions influence daily variations of NO, $NO_2$ and $O_3$.] Consequently, the average annual variation of $O_3$, together with that of Ox, has the greatest values in the summer (June and July) (Fig. 4) (Makra et al, 2001a). In Szeged, the average annual variation of $PM_{10}$ has the greatest values in November, December and January and the lowest ones in the summer months (Fig. 8). Stable anticyclonic weather situations favour accumulation of pollutants. The lowest values during the summer (June, July, August and September) can be explained by dilution caused by intensive vertical exchange in the atmosphere (Fig. 8) (Makra et al., 2001a; Makra, 2005; Makra et al., 2010).

**Weekly and diurnal variations of the pollutant concentrations**

The diurnal variations of NO and $NO_2$ have the shape of a double wave, with bigger amplitudes for NO than for $NO_2$ (Fig. 9a). Due to the traffic density, the concentration of NO is relatively higher on weekdays than on weekends. This effect can also be observed for the secondary substance $NO_2$. The average diurnal variations on weekdays are greater for NO than for $NO_2$, because $NO_2$ has a longer lifetime than the more reactive NO (Fig. 9a). Generally, the NO concentrations are higher in the morning than in the evening. This can be explained by the fact that in the morning the rush hour is shorter, and the atmosphere near the surface is more stable than in the evening. The low NO concentrations early in the afternoon result mainly from the reduction of $O_3$ by NO. A maximum takes place in the early afternoon caused by photochemical $O_3$ formation, while a minimum occurs after midnight. On the basis of its definition, the diurnal variation of Ox is similar to that of $O_3$ (Fig. 9a). On weekends the average $O_3$ maximum values are higher than on weekdays, but this is not valid for Ox (Makra et al., 2001a; 2010).
The weekly and diurnal variations of $PM_{10}$ have the shapes of double waves (Fig. 9b). Both primary and secondary maxima can be observed during peak hours and, in the same way, primary and secondary minima occur when traffic is the lowest (at night) or is decreasing (around midday). Also, due to the dense traffic, the concentrations of $PM_{10}$ is relatively higher on weekdays and lower at weekends (Fig. 9b).

**Air quality of Szeged based on different indices**

Standards for the assessment of single air pollutants exist in almost every country of the world, e.g. in EU directives. However, these standards are insufficient in view of the persistent demands (e.g. from planners) for the assessment of air quality as a whole, which is not limited to a single air pollutant. Therefore, indices on the basis of routinely monitored air pollutants were developed. The indices represented here are only statistical and have no direct relation to the well-being and health of human beings. They indicate mainly the content of air pollution in the ambient air and are therefore called air stress indices ASI (Mayer et al., 2002b). They can be calculated according to the following formulae:
Planning-related air stress index $ASI_1$ for mean stress, developed by the Office of Environmental Protection, Division of Urban Climate, City of Stuttgart, Germany:

$$ASI_1 = \frac{1}{4} * \left( \frac{C(SO_2)}{20\,\mu g\,m^{-3}} + \frac{C(NO_2)}{40\,\mu g\,m^{-3}} + \frac{C(PM_{10})}{40\,\mu g\,m^{-3}} + \frac{C(benzene)}{5\,\mu g\,m^{-3}} \right),$$

C: arithmetical annual mean values ($\mu g\ m^{-3}$); reference values (denominators of the addable sums): air pollutant specific EU standards (Mayer et al., 2002b).

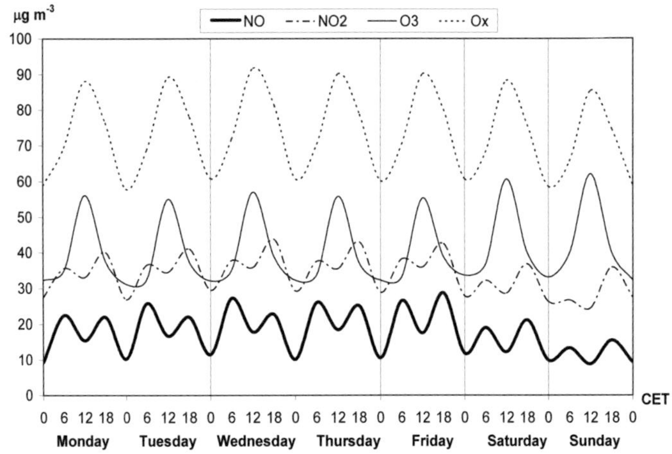


(a)

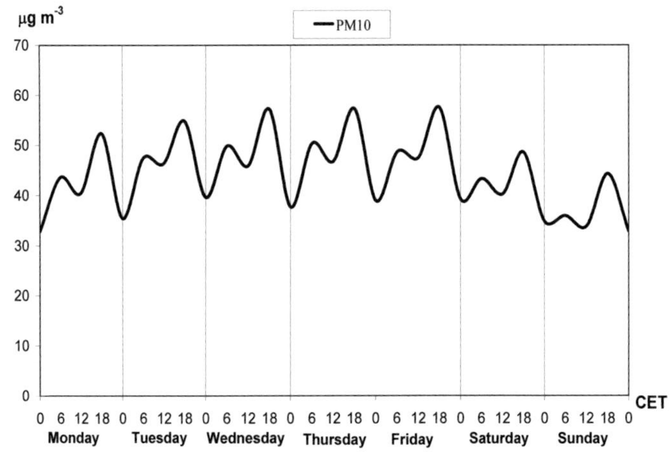


(b)

Fig. 9. Average weekly and diurnal variations of NO, $NO_2$, $O_3$, Ox (a) and $PM_{10}$ (b), monitoring station, Szeged, 1997-2001

Planning-related air stress index $ASI_2$ for short-term stress, developed by the Office of Environmental Protection, Division of Urban Climate, City of Stuttgart, Germany:

$$\mathrm{ASI}_2 = \frac{1}{4} * \left( \frac{N(\mathrm{SO}_2)}{24} + \frac{N(\mathrm{NO}_2)}{18} + \frac{N(\mathrm{PM}_{10})}{35} + \frac{N(\mathrm{CO})}{1} \right),$$

*N*: number of cases per calendar year when air pollutant specific EU limit values are exceeded [$SO_2$: 350 µg m$^{-3}$ (1 h mean value), $NO_2$: 200 µg m$^{-3}$ (1 h mean value), $PM_{10}$: 50 µg m$^{-3}$ (daily mean value), CO: 10 mg m$^{-3}$ (highest daily running 8 h mean value)] (Mayer et al., 2002b).

A graded assessment scale (Table 3) is available for the air stress indices $ASI_1$ and $ASI_2$ (Mayer et al., 2002b), which can for example serve as a basis for planning specific recommendations with respect to air quality.

| | | |
|---|---|---|
| *$ASI_1$: no single air pollutant exceeds the corresponding limit value* | | |
| *$ASI_2$: no single air pollutant shows a higher number of cases per calendar year when air pollutant specific limit values are exceeded* | | |
| level I | very low air stress | $ASI_1$, $ASI_2 < 0.2$ |
| level II | low air stress | $0.2 \le ASI_1$, $ASI_2 < 0.4$ |
| level III | moderate air stress | $0.4 \le ASI_1$, $ASI_2 < 0.6$ |
| level IV | distinct air stress | $0.6 \le ASI_1$, $ASI_2 < 0.8$ |
| level V | strong air stress | $ASI_1$, $ASI_2 \ge 0.8$ |
| *$ASI_1$: not less than one air pollutant exceeds the corresponding limit value* | | |
| *$ASI_2$: not less than one air pollutant shows a higher number of cases per calendar year when air pollutant specific limit values are exceeded* | | |
| level VI | extreme air stress | independent of $ASI_1$ and $ASI_2$ |

Table 3. Assessment of air quality conditions on the basis of $ASI_1$ and $ASI_2$ (Mayer et al., 2002b)

Considering only values of either $ASI_1$, or $ASI_2$, the air quality of Szeged can be characterized by strong air stress (level V) in each examined year. For further analysis (Table 3): on the one hand, concentration of $PM_{10}$ (considered for calculation of $ASI_1$) exceeds its limit value in each of the five years; on the other hand, both for $PM_{10}$ and CO the number of actual times specific limit values were exceeded (considered for calculation of $ASI_2$) is several times higher than that of the limits in all five years. Consequently, independently of the actual values of either $ASI_1$ or $ASI_2$, the air quality of Szeged can be characterized by extreme air stress (level VI) (Table 3).

Air stress index $ASI_{Sz}$ on a daily basis, developed by the Federal State Institute for Environmental Protection Baden-Wuerttemberg, Karlsruhe, Germany:

$$ASI_{Sz} = \frac{C(SO_2)}{350\,\mu g\,m^{-3}} + \frac{C(CO)}{10\,mg\,m^{-3}} + \frac{C(NO_2)}{200\,\mu g\,m^{-3}} + \frac{C(O_3)}{180\,\mu g\,m^{-3}} + \frac{C(PM_{10})}{50\,\mu g\,m^{-3}}.$$

Lower index $_{Sz}$ indicates datasets of Szeged, to which this air stress index is applied. $C(SO_2)$, $C(NO_2)$, and $C(O_3)$: highest daily 1 h mean values ($\mu g\ m^{-3}$), $C(CO)$: highest daily running 8 h mean value ($mg\ m^{-3}$), $C(PM_{10})$: daily mean value ($\mu g\ m^{-3}$); limit values based on EU directives (Mayer et al., 2002b).

$ASI_{Sz}$ classes and ranges are as follows: I: $ASI_{Sz} < 0.5$; II: $0.5 \le ASI_{Sz} < 1.1$; III: $1.1 \le ASI_{Sz} < 1.7$; IV: $1.7 \le ASI_{Sz} < 2.3$; V: $2.3 \le ASI_{Sz} < 2.9$; VI: $ASI_{Sz} \ge 2.9$.

Impact-related indices, which are called air quality indices, constitute the second group of indices for the assessment of the air quality effective complex. Such indices are very rare, because it is difficult to quantify the impacts of air pollutants on the well-being and health of human beings. The methodology of air quality indices is to assign concentrations of ambient air pollutants to different air pollutant specific ranges. *The air quality index itself is represented by the highest index class among the considered air pollutants.* The relation to the impact on human beings is given by different classified ranges of air pollutant concentrations, which are derived from epidemiological and toxicological investigations (Mayer et al., 2004).

A new impact-related air quality index obtained on a daily basis and abbreviated as DAQx (**D**aily **A**ir **Q**uality Inde**x**) was recently developed and tested by the Meteorological Institute,

University of Freiburg, and the Research and Advisory Institute for Hazardous Substances, Freiburg, Germany (Mayer et al., 2002a; 2002b). DAQx considers the air pollutants $SO_2$, CO, $NO_2$, $O_3$, and $PM_{10}$. To enable a linear interpolation between index classes, DAQx is calculated for each air pollutant (x) by

$$\mathrm{DAQx} = \left[ \left( \frac{\mathrm{DAQx_{up}} - \mathrm{DAQx_{low}}}{C_{up} - C_{low}} \right) * (C_{inst} - C_{low}) \right] + \mathrm{DAQx_{low}} ,$$

with $C_{inst}$: highest daily 1 h concentration of $SO_2$, $NO_2$, and $O_3$, highest daily running 8 h mean concentration of CO, and mean daily concentration of $PM_{10}$; $C_{up}$: upper threshold of specific air pollutant concentration range; $C_{low}$: lower threshold of specific air pollutant concentration range; $\mathrm{DAQx_{up}}$: index value according to $C_{up}$; $\mathrm{DAQx_{low}}$: index value according to $C_{low}$ (Mayer et al., 2004).

| $SO_2$ ($\mu g\ m^{-3}$) | CO ($mg\ m^{-3}$) | $NO_2$ ($\mu g\ m^{-3}$) | $O_3$ ($\mu g\ m^{-3}$) | $PM_{10}$ ($\mu g\ m^{-3}$) | DAQx value | DAQx class | *Classification* |
|---|---|---|---|---|---|---|---|
| 0– 24 | 0.0– 0.9 | 0– 24 | 0– 32 | 0.0– 9.9 | 0.5–1.4 | 1 | *very good* |
| 25– 49 | 1.0– 1.9 | 25– 49 | 33– 64 | 10.0–19.9 | 1.5–2.4 | 2 | *good* |
| 50– 119 | 2.0– 3.9 | 50– 99 | 65–119 | 20.0–34.9 | 2.5–3.4 | 3 | *satisfactory* |
| 120– 349 | 4.0– 9.9 | 100–199 | 120–179 | 35.0–49.9 | 3.5–4.4 | 4 | *sufficient* |
| 350– 999 | 10.0–29.9 | 200–499 | 180–239 | 50.0–99.9 | 4.5–5.4 | 5 | *poor* |
| $\geq 1000$ | $\geq 30.0$ | $\geq 500$ | $\geq 240$ | $\geq 100$ | $\geq 5.5$ | 6 | *very poor* |

Table 4. Assignment of ranges of specific air pollutant concentrations to DAQx values and DAQx classes inclusive of classification names according to school marks (Mayer et al., 2002a; 2002b; 2004)

To investigate the sensitivity of indices for the assessment of air quality conditions, frequency distributions for $ASI_{Sz}$ as an exponent of air stress indices and DAQx as an exponent of air quality indices were calculated for Szeged downtown. $ASI_{Sz}$ as well as DAQx are indices on a daily basis. Since $ASI_{Sz}$ has no relation to the impact on human beings, six classes were statistically defined on the results of five-year (1997-2001) daily values (Table 4) (Makra et al., 2003).

The daily values of both $ASI_{Sz}$ air stress index and DAQx air quality index were calculated for the examined five-year period. Results are shown only for 2001 (Figs. 10-11). Empty sections on the figures indicate lack of data. $ASI_{Sz}$ values - exceeding level III - presenting increased air stress as well as peak values are concentrated in the winter half-year or in the winter months (Fig. 10). This can be explained by climatic reasons. The standard deviation of DAQx values is less than that of $ASI_{Sz}$ values. Peak values of DAQx are also concentrated in the winter half-year (Fig. 11); however, this is not as characteristic as in the case of $ASI_{Sz}$ values. The reason of this is that the vertical axis of the diagram for $ASI_{Sz}$ is linear, while that of the diagram for DAQx is not (Makra et al., 2003).

Frequency distribution of $ASI_{Sz}$ and DAQx according to both classes and years are different. DAQx values indicate generally higher frequencies in levels IV and V; on the other hand, they are sparse in the remaining levels compared to the frequency distribution of $ASI_{Sz}$ values in levels I-VI (Fig. 12) (Makra et al., 2003; Mayer et al., 2004).

CO and $PM_{10}$ are mainly responsible for the frequency changes of DAQx classes (Fig. 13) (Makra et al., 2003; Mayer et al., 2004).

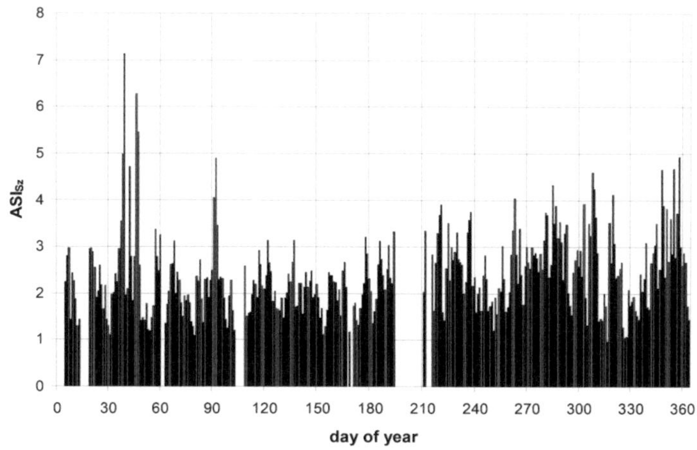


Fig. 10. Air stress index ASI for Szeged ($ASI_{Sz}$) on a daily basis, for the days of year 2001

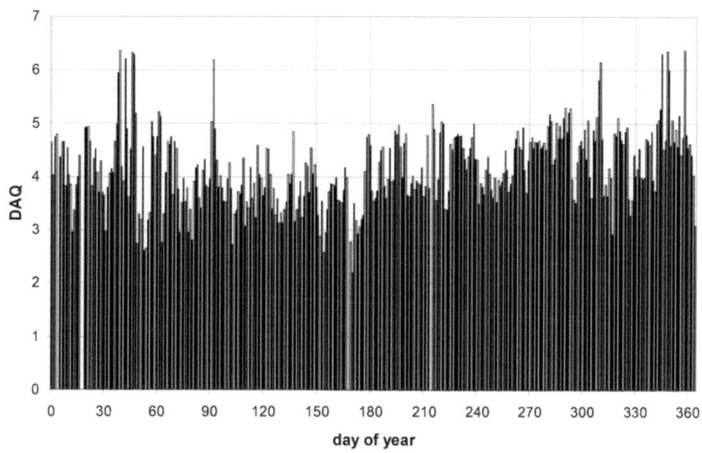


Fig. 11. Air quality index highest DAQx for Szeged on a daily basis, for the days of year 2001

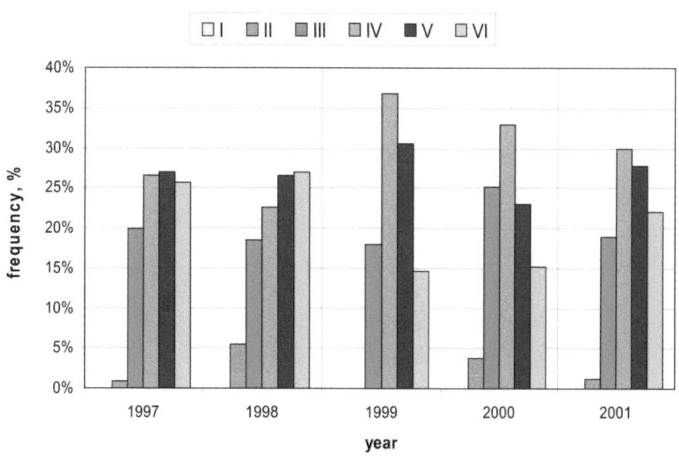


Fig. 12. Frequency distribution of $ASI_{Sz}$ for Szeged, 1997–2001

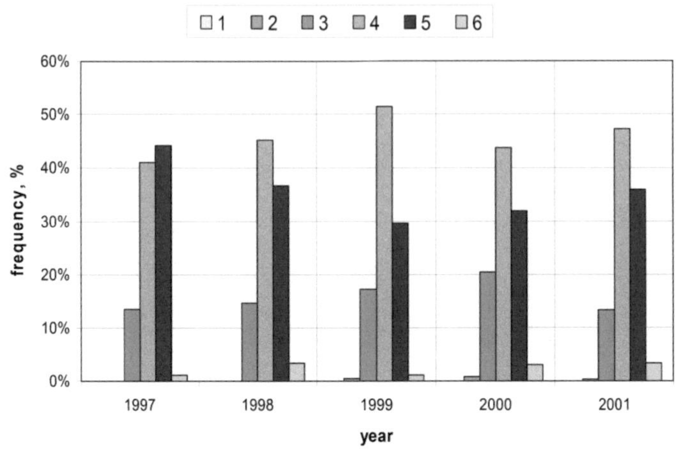


Fig. 13. Frequency distribution of highest daily DAQx for Szeged, 1997-2001

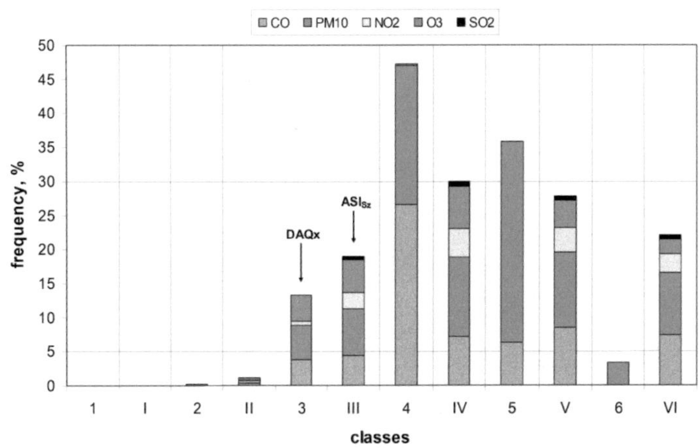


Fig. 14. Frequency distribution of $ASI_{Sz}$ and highest daily DAQx values according to classes, with the ratio of the pollutants, 1997-2001

Share of the different pollutants in $ASI_{Sz}$ and DAQx values (Fig. 14) can be clearly deduced by the formula of $ASI_{Sz}$ and the DAQx classes (Table 4), respectively. Analysis of both $ASI_{Sz}$ and DAQx values represents high pollution load of $PM_{10}$ and carbon-monoxide. $PM_{10}$ and CO several times exceeded their standards, substantially modifying the air quality of Szeged (Fig. 14).

### Statistical relationship between air pollution and meteorological parameters

This section deals with the levels of air pollutants in Szeged in relation to Péczely's weather types based on sea level pressure, subjectively defined over the Carpathian Basin (Péczely, 1957). Specific large-scale weather situations both for winter and summer months were found to play a significant role in the pollutant concentrations in Szeged. Since no ozone parameters show significant Péczely's inter-weather type differences in mean concentrations neither in the summer nor winter months, the secondary pollutants were omitted from further consideration.

On the basis of the mean values of the air pollutants for the days belonging to the 13 Péczely types and Péczely's weather type - air pollution difference matrices, the following characteristics of their inter-relationships are observed. The short interpretation of the 13 Péczely's large-scale weather situations are indicated below Table 5.[3] In the winter, types 9, 12 and 11 are the most efficient in enriching the air pollutants, while (in decreasing order) types 11, 6 and 1 in diluting them. The dubious role of type 11 can probably be attributed to the fact that on the days belonging to this type substantially different wind speeds can occur. In the summer, the role of type 5 is exclusive (78.1%) in enriching the pollutants, at the same time types 2, 8, 1 and 11 are the most important in diluting them (Table 5).
During the study period, the number of industries around and inside the city, as well as the number of cars, did not change substantially and therefore the emissions can be considered stable. Thus, the findings of this section record the influence of the atmospheric circulation in the air quality of Szeged relatively well. This would not be the case if some large industrial units had started or ceased operation in the neighbouring countries and thus the long-range transport of pollutants had modified the pre-existing atmospheric situation in Szeged; however, such great changes, according to our knowledge, did not occur during the study period.

| Péczely-type | levels of the air pollutants in the pairwise comparisons of the Péczely-types are | | | | | | | |
|---|---|---|---|---|---|---|---|---|
| | [1] enriched | | | | [2] diluted | | | |
| | ranking | | % | | Ranking | | % | |
| | winter | summer | winter | summer | winter | summer | winter | summer |
| 1 (mCc) | 6–10 | 2–3 | 2.0 | 6.3 | 3 | 3–4 | 13.2 | 12.9 |
| 2 (AB) | 6–10 | 7–13 | 2.0 | 0.0 | 4–5 | 1 | 11.3 | 35.5 |
| 3 (CMc) | 12–13 | 7–13 | 0.0 | 0.0 | 12–13 | 7–9 | 0.0 | 3.2 |
| 4 (mCw) | 6–10 | 7–13 | 2.0 | 0.0 | 4–5 | 7–9 | 11.3 | 3.2 |
| 5 (Ae) | 5 | 1 | 7.8 | 78.1 | 11 | 10–13 | 1.9 | 0.0 |
| 6 (CMw) | 4 | 4–6 | 9.8 | 3.1 | 2 | 5–6 | 17.0 | 6.5 |
| 7 (zC) | 12–13 | 7–13 | 0.0 | 0.0 | 7 | 10–13 | 5.7 | 0.0 |
| 8 (Aw) | 6–10 | 7–13 | 2.0 | 0.0 | 8–10 | 2 | 3.8 | 16.1 |
| 9 (As) | 1 | 4–6 | 31.4 | 3.1 | 8–10 | 10–13 | 3.8 | 0.0 |
| 10 (An) | 6–10 | 4–6 | 2.0 | 3.1 | 6 | 10–13 | 7.5 | 0.0 |
| 11 (AF) | 3 | 7–13 | 13.7 | 0.0 | 1 | 3–4 | 20.7 | 12.9 |
| 12 (A) | 2 | 2–3 | 25.5 | 6.3 | 8–10 | 7–9 | 3.8 | 3.2 |
| 13 (C) | 11 | 7–13 | 1.8 | 0.0 | 12–13 | 5–6 | 0.0 | 6.5 |

Table 5. Ranking and ratio of the Péczely's weather types in enrichment/dilution of the air pollutants, %, air pollutants are considered together, 1997-2001
[1](1 = most enriched; 13 = least enriched)
[2](1 = most diluted; 13 = least diluted)
[3]Type 1 (mCc): Hungary lies in the rear part of an East-European cyclone; Type 2 (AB): Anticyclone over the British Isles; Type 3 (CMc): Hungary lies in the rear part of a Mediterranean cyclone; Type 4 (mCw): Hungary lies in the fore part of a West-European cyclone; Type 5 (Ae): Anticyclone east of Hungary; Type 6 (CMw): Hungary lies in the fore part of a Mediterranean cyclone; Type 7 (zC): Zonal, cyclonic; Type 8 (Aw): Anticyclone extending from the west; Type 9 (As): Anticyclone south of Hungary; Type 10 (An): Anticyclone north of Hungary; Type 11 (AF): Anticyclone over the Fennoscandinavian region; Type 12 (A): Anticyclone over the Carpathian Basin; Type 13 (C): Cyclone over the Carpathian Basin.

On the basis of previous works, connection of both the objectively determined air-mass types (Makra et al., 2006a; 2006b) and the subjectively defined Péczely's weather types (Makra et al., 2007a; 2007b) on the one hand, and pollen grain and chemical pollutants concentrations in Szeged, on the other, detected that pollen and chemical pollutant levels can be associated to different pressure patterns ruling the region examined.

When considering pollen release of the species considered in the pollination period, some objective and subjective weather types are favourable, while others are negligible in the classification of pollen levels. On the other hand, when analysing levels of the chemical pollutants: (1) objective types with anticyclonic character are mostly favourable both in winter and summer, while those with cyclonic character are mostly negligible in winter (Makra et al., 2006a); (2) Péczely's anticyclonic types in winter are mostly favourable, while cyclonic ones are mostly negligible in classification of pollutant levels; at the same time, in the summer none of them is predominant (Makra et al., 2007a). Hence, while the objective weather types have a significant role, Péczely's large-scale weather situations cannot be considered as an overall system in the categorization of pollutants concentrations.

Accordingly, for the classification of pollen release of the species, neither the cyclonic nor the anticyclonic weather types show a clear character within the objective and the subjective weather classification systems. On the other hand, for classifying the chemical pollutants, the objective types with anticyclonic character are effective in both extreme seasons (Makra et al., 2006a), while Péczely's cyclonic and anticyclonic types have an emphasized role only in winter, whereas they are inefficient in summer (Makra et al., 2007a).

When disregarding anticyclonic and cyclonic character of the weather types and taking into account pairwise comparisons for each type, efficiency of the pollen related objective air mass types seems to be much higher than that of Péczely's weather types. Furthermore, the efficiency of the chemical pollutants related objective types seems to be significant in both seasons, while a substantial decrease of that can be observed for Péczely's classification in the summer months. Hence, Péczely types seem practically useless in classifying air pollutants in summer. As a result of the above, the objective air mass types are more efficient than Péczely's weather types in classifying either biological or chemical air pollutants and in addition can be used more efficiently in the air pollution forecast (Makra et al., 2009).

## 7. The effect of traffic on air quality by applying CAR and CALINE4 dispersion models

EU urban areas host 80% of the population and generate 75 to 85% of the gross national product. They play an essential role in the vitality and competitiveness of Europe. However, this vitality is today endangered by the impact of non-sustainable urban modes of transport. The use of private cars generates pollution, high energy consumption, noise, congestion and accidents. Reducing emissions, improving air quality, reducing accidents and congestion, reducing social deprivation, and thus increasing the quality of life in urban areas, requires a modal shift from private transport to public and sustainable transport. The aim of this section is to analyse how air pollution of vehicular traffic changed on the main roads of the Szeged region (Fig. 6, 21) in the period 1995–2008.

Air pollutant concentrations were simulated using CALINE4 and CAR road traffic dispersion models. The spatial and temporal resolutions of both models are indicated in Fig 5. The main characteristics of the CAR model in comparison with those of the CALINE4 model are summarized in Table 6. For a more detailed description see Gyöngyösi, et al (2009).

| | CAR | CALINE4 |
|---|---|---|
| Type of application | Air quality assessment<br>Regulatory purposes and compliance<br>Policy support | Air quality assessment |
| Model output | Concentration | Concentration |
| Type air pollution source | Traffic emissions (line source) | Traffic emissions (line source) |
| Release type | Continuous release without interruption | Continuous release without interruption |
| Spatial scale of model | Local (up to 30 km) | Local (up to 30 km) |
| Form of release | Carbon monoxide (CO)<br>Nitrogen oxides ($NO_x$)<br>Lead (Pb) | Carbon monoxide (CO)<br>Nitrogen oxide ($NO_x$)<br>benzene (VOCs and TSP) |
| Type of model | statistical | Gaussian |
| Duration of validity | annual average | annual average<br>1 to 24 hours percentiles |
| Computer platform | PC | PC |

Table 6. Comparison of the main characteristics of the CALINE4 and CAR road traffic models

### 7.1 Description of the CALINE4 model

CALINE4 is a fourth generation line source air quality dispersion model developed by the California Department of Transportation. It is based on "Gaussian" and "Fickian" diffusion equations and employs a mixing zone concept to characterize pollutant dispersion over the roadway. A fairly extensive evaluation of the line source dispersion model CALINE4 was performed by Benson (1992).

The purpose of the model is to assess air quality impacts near transportation facilities, in other words to predict air pollution concentrations near roadways. Given source strength, meteorology and site geometry, CALINE4 can reliably predict pollutant concentrations for a receptor located within 150 metres of roadways and can be used with reliability until within 500 m of roadways. In addition to predicting concentrations of relatively inert pollutants such as carbon monoxide (CO), the model can predict nitrogen dioxide ($NO_2$) and suspended particle concentrations. It also has special options for modelling air quality near intersections, street canyons and parking facilities.

The equation below shows the Gaussian diffusion equation for line sources, including reflection, for a differential highway length with uniform source strength.

$$\mathrm{d}C = \frac{q \cdot \mathrm{d}y}{2\pi u \sigma_y \sigma_z}\left(\exp\left(\frac{-y^2}{2\sigma_y^2}\right)\right)\left\{\exp\left(\frac{-(z-H)^2}{2\sigma_z^2}\right) + \exp\left(\frac{-(z+H)^2}{2\sigma_z^2}\right)\right\},$$

where:

dC: incremental concentration (μg $m^{-3}$),

$q$: Uniform lineal source strength (μg s$^{-1}$ m$^{-1}$),
d$y$: differential highway length with uniform source strength (m),
$u$: wind speed (m s$^{-1}$),
$z$: receptor height (m),
$H$: source height (m),
$\sigma_y$, $\sigma_z$: are horizontal and vertical dispersion parameters, respectively (m).

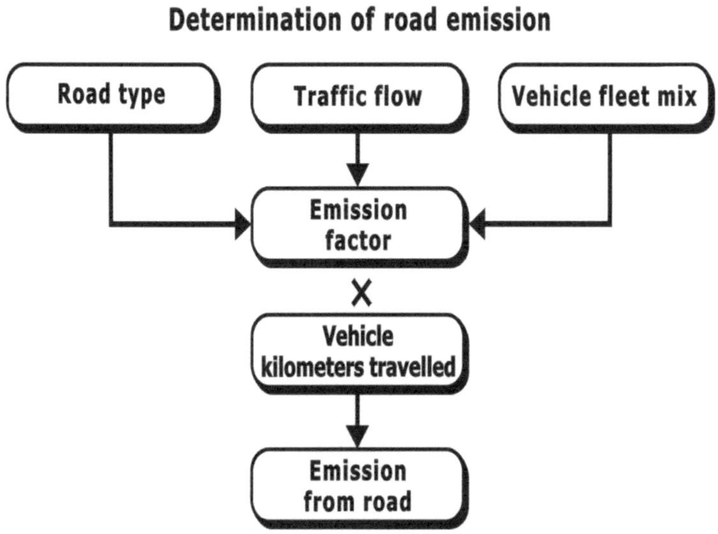


Fig. 15. Calculation of emissions in the CALINE4 model (Benson, 1992)

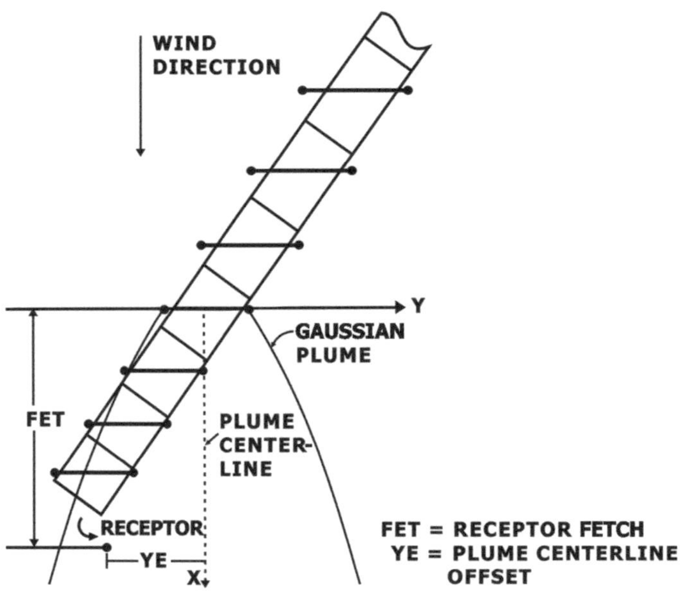


Fig. 16. Application of a Gaussian plume for road segments (Benson, 1992)

| Parameters | Unit |
|---|---|
| **I. Meteorological parameters** | |
| air temperature | °C |
| wind speed | m $s^{-1}$ |
| wind direction | degree |
| standard deviation of the wind direction | degree |
| atmospheric stability | A - G |
| height of the mixing layer | m |
| aerodynamic roughness | cm |
| **II. Characteristics of the road** | |
| traffic density | vehicle / hour |
| emission factor | g / vehicle mile |
| height of the road | m |
| width of the road | m |
| **III. Position of the receptors** | |
| x coordinate | m |
| y coordinate | m |
| z coordinate | m |

Table 7. Input parametres of CALINE4 model

CALINE4 divides individual highway links into a series of elements from which incremental concentrations are computed and then summed to form a total concentration estimate for a particular receptor location (Figs. 15–16). Each segment of a roadway link is treated as a separate emission source producing a plume of pollutants which disperses downwind. Pollutant concentrations at any specific location are calculated using the total contribution from overlapping pollution plumes originating from the sequence of roadway segments (Figs. 15–16).

Downwind concentrations from the element are modelled using the crosswind FLS (Finite Line Source) Gaussian formulation, but $\sigma_y$ and $\sigma_z$ are modified to consider the mechanical turbulence created by moving vehicles and the thermal turbulence generated by hot vehicle exhaust over the highway (Fig. 17).

CALINE4 treats the region directly over the highway as a zone of uniform emissions and turbulence. This is designated as the mixing zone (Fig. 17), and is defined as the region over the travelled way (traffic lanes not including shoulders) plus three metres on either side. The additional width accounts for the initial horizontal dispersion imparted to pollutants by the vehicle wake. Within the mixing zone, the mechanical turbulence created by moving vehicles and the thermal turbulence created by hot vehicle exhaust are assumed to be the dominant depressive mechanisms (Fig. 17). In CALINE4, a computational scheme called the Discrete Parcel Method (Benson, 1984) is used to model $NO_2$ concentrations. The Discrete Parcel Method fixes the initial mixing zone concentrations of the reactants on the basis of

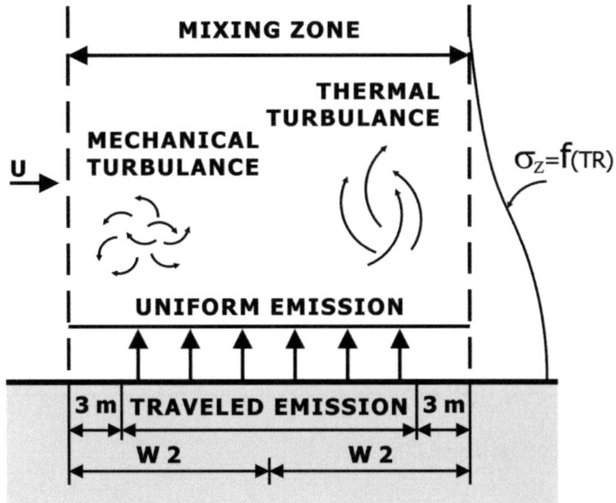


Fig. 17. Mixing zone concept in CALINE4 (Benson, 1992)

ambient and vehicular contributions. Input values required by CALINE4 to compute the pollutant dispersion include: link geometry (highway), link activity (traffic, emission), run conditions and receptor locations. Table 7 shows a complete list of all input parameters. Furthermore, Table 8 is intended to help the user assign proper and realistic values to the input variables used by the model. The limits of different CALINE4 input variables are also involved.

Input requirements

The Gaussian formulation used is based on two somewhat restrictive assumptions: 1) horizontally homogeneous wind flow; 2) steady-state meteorological conditions. For these reasons CALINE4 needs a simple orography.
There are some drawbacks, however, to this model: Since it is a Gaussian dispersion model and uses modified Gaussian dispersion parameters ($\sigma_y$ and $\sigma_z$) which are fairly accurate only up to a distance of 10 km, it is important that no road links should exceed 10 km in length. The model also specifies that in a single run, the total number of links should not be more than 20. A required parameter from the run options is wind speed. The model can handle only wind speeds exceeding 0.5 m s$^{-1}$. For each receptor (maximum number of receptors is 20), inputs are required for coordinates x, y and height (z). The user must also define CO emission factors for each roadway link.

### 7.2 Results from the CALINE4 model

Annual mean concentrations, as means of all receptors for both CO (Fig. 18a) and $NO_2$ (Fig. 18b) on the major roads of the Szeged region modelled at a height of one metre, show clearly increasing trends. Annual mean CO and $NO_2$ concentrations on some major roads of the Szeged region, considered on a function of the receptors, indicate higher values in the downtown and lower ones towards the outskirts. Furthermore, they show partly slight (CO) and partly definite ($NO_2$) increasing trends (Figs. 18a–18b). Annual mean CO (Figs. 19a–19b) and $NO_2$ (Figs. 20a–20b) concentrations on the major roads of Szeged region are calculated using the CALINE4 model for each year in the period 1995–2007. However, CO and $NO_2$

| Variable | Suggested / mandatory limits | Comments |
|---|---|---|
| Surface Roughness ($z_0$) | $3 \leq z_0 \leq 400$ cm | 15% of average canopy height |
| Settling Velocity ($Vs$) | $Vs \geq 0$ | $Vs = 2.98 \cdot 10^5 \cdot D_{ar}^2$ ; $D_{ar}^2$ = aerodynamic resistance diameter (cm) |
| Deposition Velocity ($Vd$) | $Vd \geq 0$ | |
| Intersection Traffic Parameters (Various) | | Traffic assumed to flow from link endpoint 1 to link endpoint 2. |
| Wind Speed ($V$) | $V \geq 0.5$ m s$^{-1}$ | Measure at 5 to 10 m or assume worst case. For localized sources and nearby receptors, wind speeds measured at lower elevations (5 m) desirable. For more diffuse sources and distant receptors, 10 m height more appropriate. |
| Stability Class (CLAS) | $1 \leq$ CLAS $\leq 7$ | Golder (1972), Turner (1964), or assume worst case. |
| Directional Variability (SIGTH) | 5° ≤ SIGTH ≤ 60° | Measure at 4 m to 10 m or assume worst case. |
| Mixing Height (MIXH) | MIXH ≥ 5 m (Note: MIXH ≥ 1000 m deactivates this algorithm) | $MIXH = \frac{0.185 \cdot V \cdot \kappa}{\ln(z / z_0) \cdot f}$ <br> $V$ = wind speed (m s$^{-1}$) <br> $z$ = height $V$ measured at (m) <br> $z_0$ = surface roughness (m) <br> $\kappa$ = von Kárman constant (0.35) <br> $f$ = Coriolis parameter |
| Temperature (TEMP) | | January mean minimum plus time period adjustment |
| Photolysis Rate (KR) | KR ≥ 0 | |
| Wind Direction (BRG) | 0° ≤ BRG ≤ 360° | Wind azimuth bearing measured relative to positive y-axis. |
| Mixing Zone Width (W) | W ≥ 10 m | Minimum of 1 lane plus 3 m per side (exception: parking lot link) |
| Link Length (LL) | W ≤ LL ≤ 10 km | Link length needs to be greater than or equal to the mixing zone width for proper element resolution and less than 10 km to stay within the range of validity for the vertical dispersion algorithm. |
| Source Height (H) | -10 ≤ H ≤ 10 m | Limits of verified model performance. |
| Receptor Height (Z) | Z ≥ 0 | For depressed sections Z ≥ H (where H is negative) is permitted for receptors within the section. |
| Mixing Width (MIXWR, MIXWL) | MIXWR ≥ W/2, MIXWL ≥ W/2 | An assigned value of zero is interpreted as no horizontal obstruction. Right (R) and left (L) determination made facing link endpoint 2. |

Table 8. Different CALINE4 input variables and their limits

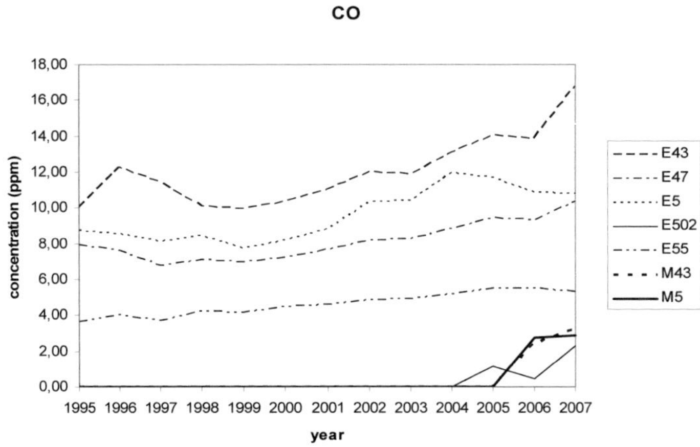


Fig. 18a. Annual mean CO concentrations in ppm (h = 1 m) , as means of all the receptors, on the major roads of the Szeged region (see the map in Fig. 21), 1 ppm CO ~ 1,15 mg $m^{-3}$ CO

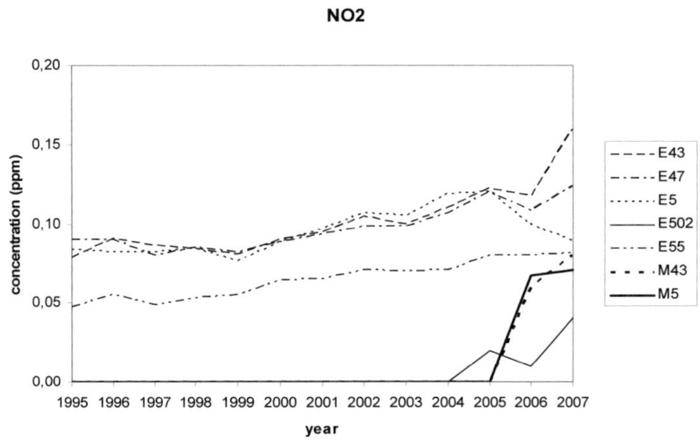


Fig. 18b. Annual mean $NO_2$ concentrations, as means of all the receptors, on the major roads of the Szeged region, ppm (h = 1 m), 1 ppm $NO_2$ ~ 1,9 mg $m^{-3}$ $NO_2$

loads for the starting year (1995) and years 2006 and 2007 are only analysed. The annual mean concentrations clearly indicate the increase of CO levels and a much more definite increase of $NO_2$ concentrations from the beginning till the end of the period examined. The difference of the concentrations is especially striking for years 2006 and 2007 (Figs. 19a–19b; Figs. 20a–20b). This is related to the highly increased transit transport through Szeged from Romania and Bulgaria after these countries entered the European Union on January 1, 2007. The vehicular traffic of the Szeged region is permanently increasing. However, in year 2007 an especially striking increase of vehicular traffic was experienced. Due to the EU membership of Romania and Bulgaria the number of trucks and lorries going through the Szeged region from these new EU countries quadrupled from 2006 to 2007. This increased traffic means a remarkable load on road no. E43. As a result, substantially higher vehicleoriginated air pollution can be experienced as clearly indicated by the CALINE4 dispersion model (Figs. 19a–19b; Figs. 20a–20b) (Makra et al., 2008).

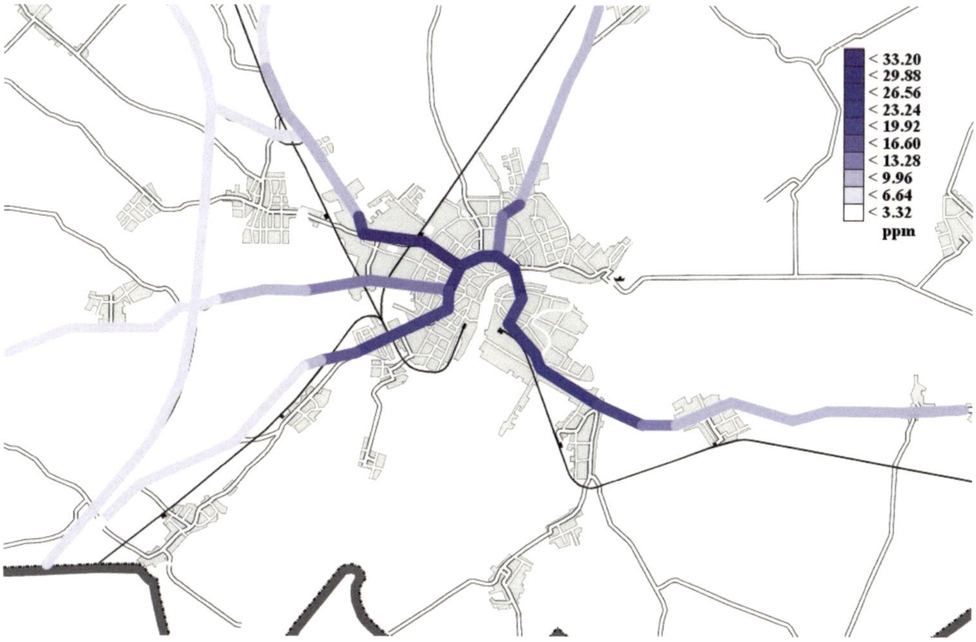


Fig. 19a. Annual mean CO concentrations on the major roads of the Szeged region, 2006, ppm (h = 1 m)

Fig. 19b. Annual mean CO concentrations on the major roads of the Szeged region, 2007, ppm (h = 1 m)

Fig. 20a. Annual mean $NO_2$ concentrations on the major roads of the Szeged region, 2006, ppm (h = 1 m)

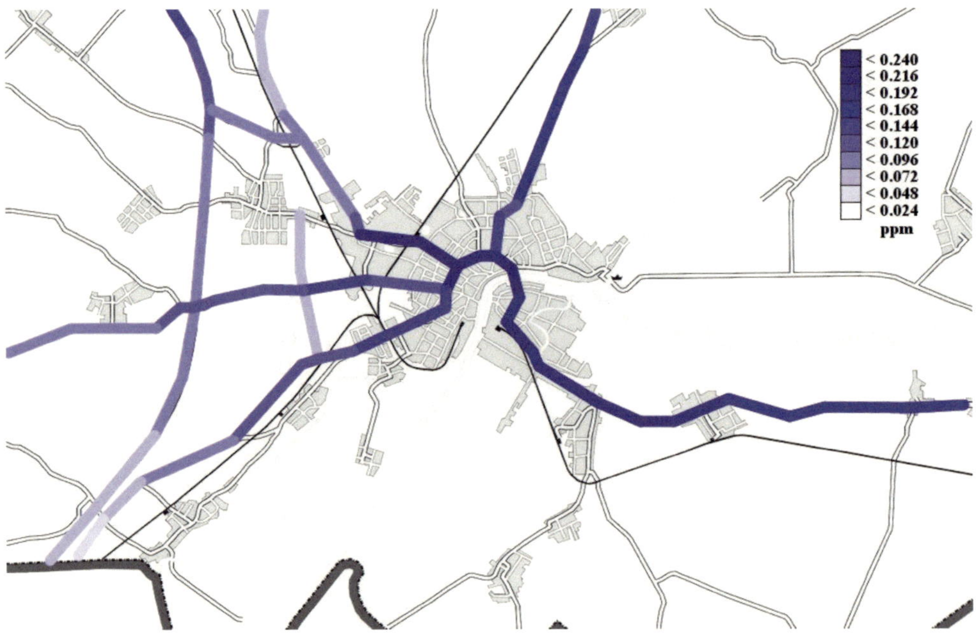


Fig. 20b. Annual mean $NO_2$ concentrations on the major roads of the Szeged region, 2007, ppm (h = 1 m)

More complex approaches to dispersion modelling are unnecessary for most applications because of the uncertainties in estimating emission factors and traffic volumes for future

years. CALINE4's accuracy is well balanced with the accuracy of state of the art predictive models for emissions and traffic. The model also possesses greater flexibility than earlier versions, at little cost to the user in terms of input complexity.
Using CALINE4 dispersion model hourly and 8-hour average concentrations can also be obtained. The hourly $NO_2$ and CO concentrations have been calculated for some main roads of Szeged near kerbsides. During worst-case meteorology $NO_2$ and CO concentrations are shown in Fig. 21. Worst-case meteorology input is the combination of the worst wind speed, wind direction, stability class. CALINE4 has special option for modelling air quality in street canyons. During this model simulation a simple form of canyon effect was also considered. Due to these facts the obtained maximum concentrations are quite high, especially in case of CO when the maximum of worst case 1-hour concentration is 10 times higher than the short-time limit values for highly protected area (~ 5 ppm).

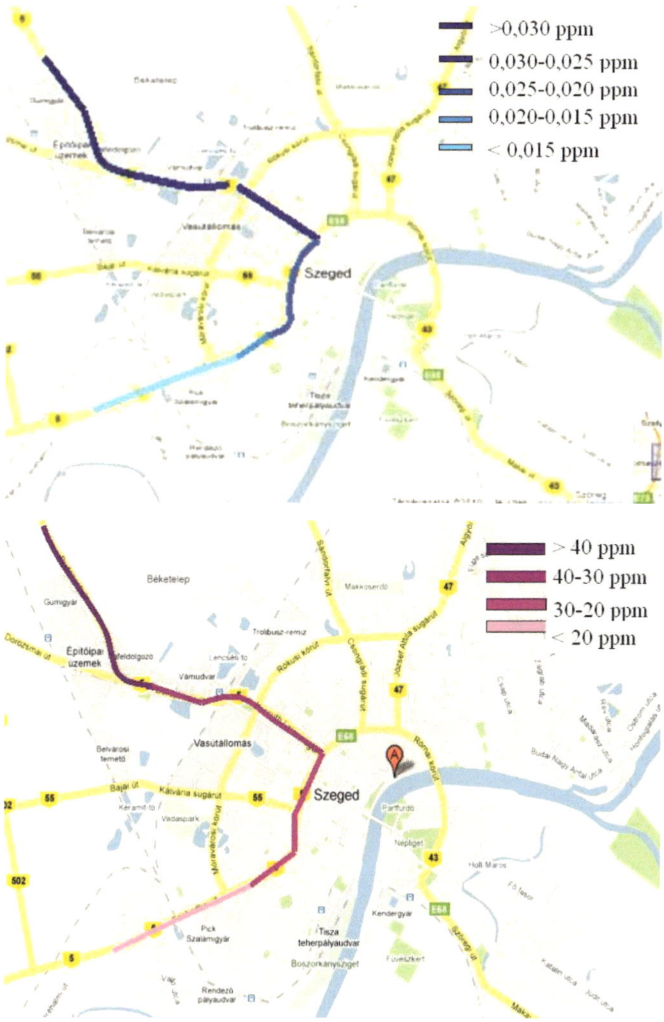


Fig. 21. Worst-case 1-hour $NO_2$ average $NO_2$ (top) and CO (bottom) concentrations at kerbsides calculated by CALINE4, 2008

In order to make a comparative calculation for year 2008, a striking change in $NO_2$ and CO concentrations is experienced along the main road no. E75, going through Szeged. Incoming traffic from Budapest direction involves smaller $NO_2$ and CO levels, respectively, compared to earlier years (Fig. 21, upper part of the coloured curves). On the other hand, outgoing traffic towards Belgrade decreased dramatically, which appeared as a substantial drop in the $NO_2$ and CO concentrations (Fig. 21, lower part of the coloured curves).
During the evaluation of different model results it was established that the calculated annual, daily and hourly mean concentration of different air pollutants can highly depend on several factors for example the distance from the road axis, the height of receptor point, the applied modelling method for averaging and parametrization procedures, applied meteorology and also the emission factors.

### 7.3 Description of the CAR model

The Dutch CAR model (Calculation of Air pollution from Road traffic) (Eerens et al., 1993) uses an empirical approach for the estimation of mean annual concentrations of $NO_2$ and non-reactive pollutants (carbon-monoxide and benzene) in urban and rural areas. The relationship between street geometry types, wind speed and concentrations of the pollutants considered were based on wind tunnel experiments.
The effect of trees along streets and the type or speed of road traffic (from highway traffic at an average speed of 100 km $h^{-1}$ to a non-continuous, stagnating traffic at an average speed of 11 km $h^{-1}$) were also considered. A source receptor function is specified for each street category as a function of distance from road axis (from 5 m to 30 m). The effect of cars and trucks are considered separately, and the resulting concentration values are calculated as a sum of the components. The model calculates emission, and regional and city background concentrations are input parameters for them. Annual averages and 1-, 8- and 24-hour 98 percentiles are the outputs of the model for each pollutant (Fig. 22).

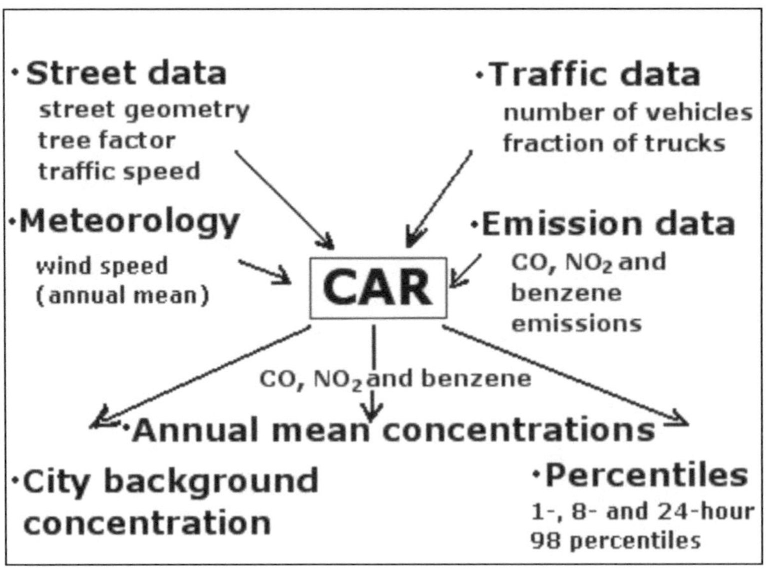


Fig. 22. Schematic diagram of system parameters, input and output data for the CAR model (Eerens et al., 1993)

### 7.4 Air quality calculations with the CAR model - input data

Input data from Szeged were used in our calculations with the CAR model. Street data, such as geometry, traffic type and vegetation have been estimated at a field trip experiment on the site. The default emission factors were changed according to the inventory of the Automotive Engineering Environmental and Energy Division at the Institute of Transport Sciences (web of the Ministry of Environment and Water, Hungary, 2009). For the exact values one may refer to Table 9 below.

Note that the units used in the CAR model (μg $m^{-1}$ $s^{-1}$ $vehicle^{-1}$) differ from those used in other sources (g $km^{-1}$). For background concentration reference, the measurements from K-puszta have been considered. For city background concentration, annual averages measured by the air quality monitoring station at Szeged were used. Concentration data collected at the air quality monitoring station were compared to the above model output.

Traffic census has been performed at 9 different sites in the city (Fig. 23; sites 1 to 9 are from top left to centre right and bottom left). Both the average daily number of motor vehicles passing through each location and the air pollution data were considered for the 12-year period 1997–2008. The temporal course of the mean daily number of vehicles at two different sites (Site 4 and 9) is also presented (Fig. 23, bottom right).

| Speed type | Speed (km $h^{-1}$) | CO g $km^{-1}$ | CO μg $m^{-1}$ $s^{-1}$ $vehicle^{-1}$ | $NO_2$ g $km^{-1}$ | $NO_2$ μg $m^{-1}$ $s^{-1}$ $vehicle^{-1}$ |
|---|---|---|---|---|---|
| | | | Cars | | |
| $V_a$ | 13 | 30.57 | 0.354 | 1.38 | 0.016 |
| $V_b$ | 22 | 21.00 | 0.243 | 1.33 | 0.015 |
| $V_c$ | 44 | 11.72 | 0.136 | 1.40 | 0.016 |
| $V_d$ | 100 | 6.40 | 0.074 | 2.45 | 0.028 |
| | | | Trucks | | |
| $V_a$ | 13 | 21.26 | 0.246 | 8.01 | 0.093 |
| $V_b$ | 22 | 15.75 | 0.182 | 6.75 | 0.078 |
| $V_c$ | 44 | 10.74 | 0.124 | 6.06 | 0.070 |
| $V_d$ | 100 | 8.86 | 0.103 | 11.28 | 0.131 |

Table 9. Emission factors for cars and trucks at different speed categories in different units. Parameters were taken from the official emission inventory of the Automotive Engineering Environmental and Energy Division at the Institute of Transport Sciences (KTI). (source: web of the Ministry of Environment and Water, Hungary; Gyöngyösi et al., 2009)

### 7.5 Air quality calculations with the CAR model - results

Results of our numerical calculations are presented in Table 10 and Fig. 24. The main findings are as follows: The measured CO concentrations showed a slight decrease, while the model yielded a 6.88 μg $m^{-3}$ $year^{-1}$ increase especially for 2008 due to higher road traffic. On the other hand, no significant change has been observed for $NO_2$ while the model yielded a gradual increase of about ~0.62 μg $m^{-3}$ $year^{-1}$. The measured data was slightly smaller than the modelled data. The model results indicate a smaller deviation from the average than the measured data (Fig. 24). The discrepancy may come partly from the fact that the emission parameters were considered constant, although the structure and technical quality of the transportation system in Szeged has been improving considerably. The concentration at the measurement site is the cumulative result of the side street traffic as well, which has changed significantly and was not treated in our calculation.

| Site | Fraction of trucks, % | traffic (number of vehicles per day) | | | CO concentration (μg m$^{-3}$); $C_b$ = 254 μg m$^{-3}$ | | |
|---|---|---|---|---|---|---|---|
| | | 12-year average | Standard deviation | Relative deviation (%) | 12-year average | Standard deviation | Relative deviation (%) |
| Site 1 | 12 | 29 150 | 8 850 | 30 | 1 010 | 220 | 22 |
| Site 2 | 9 | 26 170 | 6 050 | 23 | 1 410 | 265 | 19 |
| Site 3 | 7 | 23 110 | 3 970 | 17 | 1 265 | 170 | 14 |
| Site 4 | 5 | 21 980 | 3 625 | 16 | 2 240 | 330 | 15 |
| Site 5 | 4 | 17 415 | 2 540 | 15 | 1 830 | 230 | 12 |
| Site 6 | 4 | 11 170 | 3 285 | 29 | 1 265 | 295 | 23 |
| Site 7 | 7 | 11 910 | 1 665 | 14 | 710 | 65 | 9 |
| Site 8 | 5 | 8 700 | 1 200 | 14 | 725 | 65 | 9 |
| Site 9 | 8 | 4 875 | 1 595 | 33 | 380 | 40 | 10 |

Table 10. Model results of 12-year integration (1997–2008) at 9 sites in Szeged for CO concentrations. 12-year averages, standard deviations and relative deviations are given for traffic and CO concentrations, respectively about 5 m away from the road axis. Average fraction of trucks and city background concentrations ($C_b$) are also given.

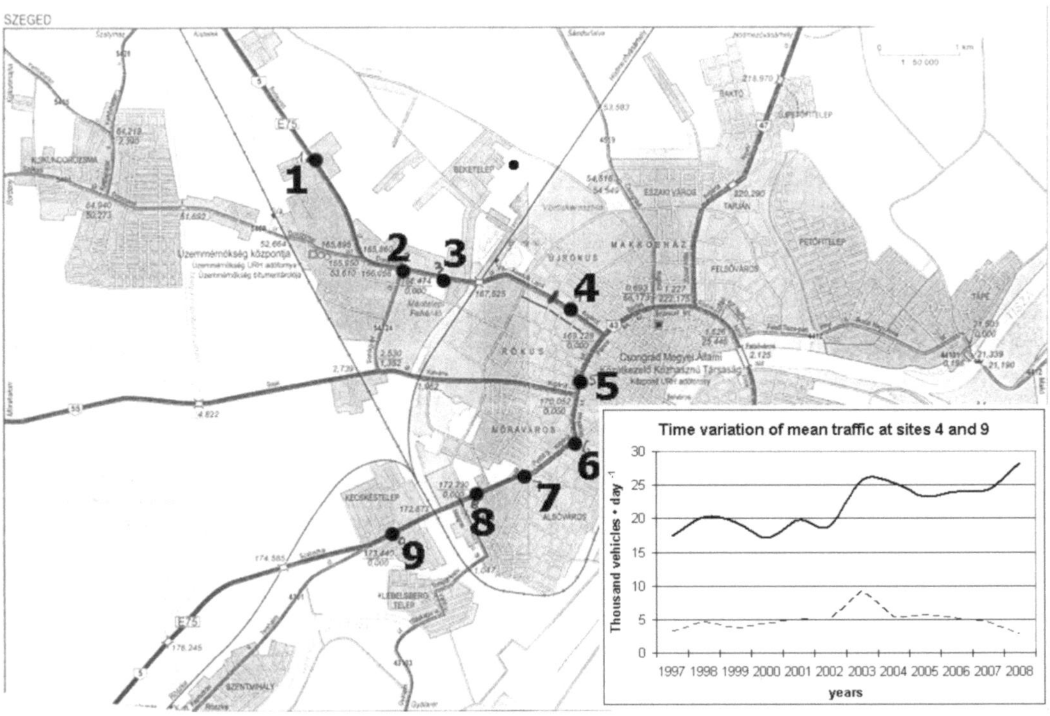


Fig. 23. Map of Szeged with the location of the measurement sites. Bottom right panel: time variation of the daily number of vehicles at two locations (site 4, solid line and site 9, dotted line) for each year (1997–2008)

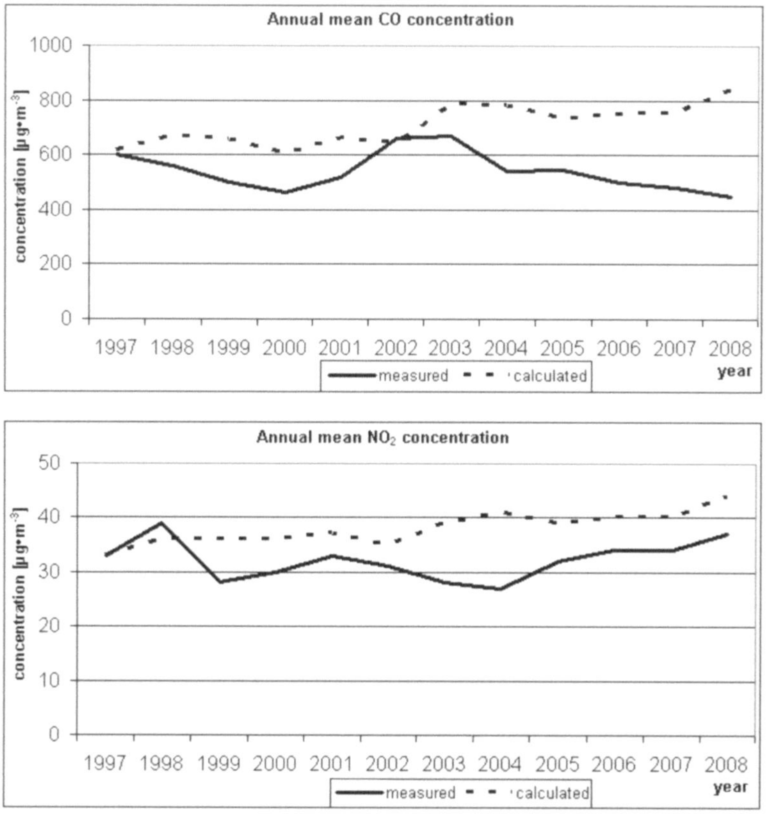


Fig. 24. Measured and calculated concentration of CO and $NO_2$ for Szeged (1997–2008) close to site 4 about 30 m away from the road axis and 3 m above the surface

An increasing trend is present for the urban average traffic for the annual means taking into account all sites (> 900 vehicles per day per year growth rate for the average during the 12-year period). Model results on two different types of urban sites are analysed: An air quality monitoring station is located at site 4, which is a typical dense urban area not so far from the city centre (with an average of 21,980 vehicles per day for the 12-year period 1997–2008), while site 9 is an open suburban site with an average of 4,875, vehicles per day.

## 8. Conclusion

In Szeged, the average annual variations of NO, $NO_2$ and $PM_{10}$ (with maxima in winter) are opposite to those of $O_3$ and Ox (with maxima in summer). The higher winter values are caused by atmospheric stability with frequent inversions. The lowest values in summer are due to dilution generated by intensive vertical exchange in the atmosphere. The highest intensities of photochemical $O_3$ formation are observed during the early afternoon in the day and the summer in the year. After removing the annual variations, the very similar average weekly variations of NO, $NO_2$ and $PM_{10}$ for Szeged show weekday maxima and weekend minima. Oppositely, those of $O_3$ show weekday minima and weekend maxima. The diurnal variations of $O_3$ show a clear daily course with one wave. On weekends the average $O_3$ maximum values are higher than on weekdays, but this is not valid for Ox . Levels of NO and $O_3$ are traffic-related. Concentrations of both CO and NO are in reverse correlation with wind speed. As both pollutants are predominantly of traffic origin, their concentrations should be correlated, which is indeed the case.

Daily average concentrations of the pollutants increased on weekdays and decreased during weekends. The concentration of $O_3$ presented an opposite trend. At the weekends, air quality improves better in the winter half-year.

Aside from single air pollutant standards, air stress indices and air quality indices enable an additional assessment of the air quality conditions, which is primarily not limited to single air pollutants. The application of air stress indices or air quality indices depends on the specific objectives of the investigation. High values of mean air stress (indicated by $ASI_1 \approx 1$) as well as extremely high values of short-term air stress (indicated by $ASI_2 > 20$) suppose a high air pollution load in Szeged. Short-term (diurnal) air pollution increased on weekdays and decreased on weekends.

Summarizing the results of the inter-relationships of Péczely's weather types and daily mean levels of the air pollutants, enrichment of the air pollutants (either in the winter or in the summer; and either considering the pollutants together or separately) occurs exclusively during Péczely's anticyclone centre or anticyclone ridge weather situations; on the other hand, their dilution can be experienced not only during cyclonic but anticyclonic ridge weather types as well. Efficiency of the objective weather classification, performed for the same period, is almost one order of magnitude higher than that of Péczely's weather types in classifying the air pollutant levels. CO, $SO_2$ and $PM_{10}$ are sensitive to the Péczely's weather classification, while $NO_2$/NO, $O_3$ and $O_{3max}$ are completely insensitive. In the winter, Péczely's anticyclonic types are mostly favourable, while Péczely's cyclonic types are mostly negligible in classification of pollutant levels. On the other hand, in the summer, none of them are predominant. Hence, although they have a clear role in the winter, Péczely's large-scale weather situations can not be considered as an overall system in categorization of pollutant concentrations, as they are inefficient in the summer.

Annual mean CO and $NO_2$ concentrations on the major roads of the Szeged region are calculated using the CALINE4 model for each year between 1995–2007. A clear increase in CO levels and a much more definite increase of $NO_2$ concentrations can be observed for the period examined. The difference of the concentrations is especially striking between years 2006 and 2007. This corresponds to the highly increased traffic through Szeged from Romania and Bulgaria after these countries entered the European Union on January 1, 2007. A comparative analysis for year 2008 resulted in a striking change in $NO_2$ and CO concentrations along the main road E75 across Szeged. Incoming traffic from Budapest involves smaller $NO_2$ and CO levels compared to earlier years, while outgoing traffic towards Belgrade radically reduced with substantial drop in both $NO_2$ and CO levels.

When studying the sensitivity of the CAR model for Szeged, the concentration of pollutants as a function of distance from road axis, the effects of wind speed, road type and tree factor on the concentration of the pollutants overall and at different traffic speeds were analysed and quantified. According to the results, much higher concentrations occur in the summer than in the winter. This is due to the fact that wind speed is the lowest from late summer until early autumn and vegetation generates a higher effect on wind speed in the summer and autumn than in the winter. Furthermore, it was detected that a doubling in the traffic (i.e. double the number of vehicles) results in a 71% increase in CO concentration. Neither the effect of trees nor the increasing traffic speed can compensate for the effect of a double truck fraction. The fraction of trucks has a great impact on $NO_2$ concentration. Summing up the results, the main findings are as follows: the level of pollution increases with (i) increasing number of vehicles, (ii) decreasing speed in urban traffic (i.e., less than 50 km $h^{-1}$), (iii) larger fraction of heavy vehicles, (iv) increasing number of trees alongside the roads and (v) smaller mean annual wind speed. In addition, the model was run on realistic input parameters, regional and city background concentrations. Model results have been compared to measurements, showing good agreement with a slight overestimation of concentration, which is due to the insufficient consideration of the technical development of the vehicles. However, modelled data show smaller deviations than measurements.

## 9. Acknowledgement

Authors would like to thank Gábor Motika for supplying data on the air pollutants and meteorological elements, Miklós Juhász for providing pollen data for Szeged, Csaba Károssy for classifying Péczely's large-scale weather situations over the Carpathian Basin on each day of the period studied, Zoltán Sümeghy for the digital mapping in Fig. 6 and Ágnes Havasi for useful comments. The contributions of László Bozó, Hungarian Meteorological Service, who kindly provided the CAR model for our study, is greatly appreciated. This study was supported by the EU-6 Projects QUANTIFY [Contract no. 003893 (GOCE)] and NitroEurope, the Hungarian Scientific Foundation (OTKA project No. PD 75500) and the Hungarian high education program TÁMOP.

## 10. References

Baklanov, A. (2006): Overview of the European project FUMAPEX. *Atmospheric Chemistry and Physics*, Vol. 6, 2005-2015, ISSN: 1680-7316

Baklanov, A.; Hänninen, O.; Slørdal, L.H.; Kukkonen, J.; Sørensen, J.H.; Bjergene, N.; Fay, B.; Finardi, S.; Hoe, S.C.; Jantunen, M.; Karppinen, A.; Rasmussen, A.; Skouloudis, A. & Sokhi, R.S. (2006) Integrated systems for forecasting urban meteorology, air pollution and population exposure. *Atmospheric Chemistry and Physics Discussion*, 6, 1867-1913, ISSN: 1680-7367

Balczó, M.; Gromke, Ch. & Ruck, B. (2009) Numerical modeling of flow and pollutant dispersion in street canyons with tree planting. *Meteorologische Zeitschrift*, Vol. 18, No. 2, 197-206, ISSN: 0941-2948

Benson, P. (1984) CALINE4 - A Dispersion Model For Predicting Air Pollutant Concentrations Near Roadways. *Report No. FHWA /CA/TL-84/15*. State of California, Department of Transportation, Division of New Technology and Research

Benson, P. (1992) A review of the development and application of the CALINE3 and 4 models. *Atmospheric Environment*, Vol. 26B, No. 3, 379-390, ISSN: 1352-2310

Benson, P.E.; Nokes, W.A. & Cramer, R.L. (1986) Evaluation of the CALINE4 line source dispersion model for complex terrain application. *Transportation Research Record*, No. 1058, 7-13, ISBN: 0-309-04052-3, Transportation Laboratory, California Department of Transportation, Sacramento, CA

Berlyand, M.E.; Burenin, N.S.; Genikhovich, E.L.; Onikul, R.I.; Panfilova, G.A. & Tsyro, S.G. (1990) Experimental investigations of atmospheric pollution due to motor vehicles. *Proceedings of the Soviet American symposium on mobile-source air pollution*, Novgorod, p. 152.

Bozó, L.; Labancz, K. & Steib, R. (2006a) Estimation of air pollution in the year 2010 based on dynamical model calculations. *Proceedings of the Scientific Days in Meteorology*, pp. 207–215, ISBN 978-963-7702-97-6, Budapest, 2006 November, Hungarian Meteorological Service, Budapest (In Hungarian)

Bozó, L.; Mészáros, E. & Molnár, Á. (2006b) *Atmospheric Environment, Modelling and Observation*, 250p, ISBN 963 05 8310 0, Akadémiai Kiadó, Budapest (In Hungarian)

Brönnimann, S. & Neu, U. (1997) Weekend-weekday differences of near-surface ozone concentrations in Switzerland for different meteorological conditions. *Atmospheric Environment*, Vol. 31, 1127-1135, ISSN: 1352-2310

Chock, D.P. (1978) A simple Line-Source Model for Dispersion Near Roadways. *Atmospheris Environment*, Vol. 12, 823-829, ISSN: 1352-2310

Eerens, H.C.; Sliggers, C.J. & Van Den Hout, K.D. (1993) The CAR model: The Dutch method to determine city street air quality. *Atmospheric Environment*, Vol. 27, No. 4, 389-399, ISSN: 1352-2310

Eskridge, R. & Catalano, J. (1987) *ROADWAY - A numerical model for predicting air pollutants near highways -user's guide*. EPA-68-02-4106, U.S. Environmental Protection Agency, Research Triangle Park, North Carolina, 125 p.

Fekete, K.; Popovics, M. & Szepesi, D. (1983) Estimation of the allowable emissions for air pollution abatement program in Hungary. *Official Issues of the Hungarian Meteorological Service*, Vol. 55, 168 p, Budapest (in Hungarian), ISSN 0133-3569

Golder, D. (1972) Relations among stability parameters in the surface layer. *Boundary-Layer Meteorology, Vol.* 3, 47-58, ISSN 0006-8314

Gyöngyösi, A.Z.; Weidinger, T.; Makra, L. & Baranka, G. (2009) Application of a dispersion model for Szeged, a medium sized Hungarian city: a case study. *Fresenius Environmental Bulletin*, Vol. 18, No. 5b, 788-797, ISSN: 10184619

Hanna, S.R. ; Egan, B.A.; Purdum, J. & Wagler, J. (2001) Evaluation of the ADMS, AERMOD and ISC3 dispersion models with the OPTEX, Duke Forest and Lovett field databases. *International Journal of Environment and Pollution*, Vol. 16, No. 1-6, 301-314, ISSN: 0957-4352

Haszpra L. (2008) EMEP – An European environmental program in the last three decades. *Légkör*, Vol. 53, No. 2, 7-12, (in Hungarian), ISSN: 0133-3666

Haszpra, L.; Barcza, Z.; Hidy, D.; Szilágyi, I.; Dlugokencky, E. & Tans, P. (2006) Trends and temporal variations of major greenhouse gases at a rural site in Central Europe. *Atmospheric Environment*, Vol. 42, 8707-8716, ISSN: 1352-2310

Havasi, A. & Zlatev, Z., 2002: Trends of Hungarian air pollution levels on a long time-scale. *Atmospheric Environment*, Vol. 36, 4145-4156, ISSN: 1352-2310

Horváth, L. (1983) Trend of nitrate and ammonium content of precipitation water in Hungary for the last 80 years. *Tellus*, Vol. 35B, No. 5, 304-308, ISSN: 0280-6509

Keuken, M.; Roemer, M. & van den Elshout, S. (2009) Trend analysis of urban $NO_2$ concentrations and the importance of direct $NO_2$ emissions versus ozone/$NO_x$ equilibrium. *Atmospheric Environment*, Vol. 43, 4780-4783, ISSN: 0894-8763

Kiss, G.; Makra, L.; Mika, J.; Borsos, E. & Motika, G., 2005: Temporal characteristics of air pollutant concentrations in Szeged, Hungary. *Acta Climatologica et Chrologica Universitatis Szegediensis*, Vol. 38-39, 125-133, ISSN: 1587-5903

Kono, H. & Ito, S. (1990) A micro-scale dispersion model for motor vehicle exhaust gas in urban areas - OMG VOLUME-SOURCE model. *Atmospheric Environment*, Vol. 24B, No. 2, 243-251, ISSN: 1352-2310

Kristóf, G.; Rácz, N. & Balogh, M. (2009) Adaptation of pressure based CFD solvers for mesoscale atmospheric problems, *Boundary-Layer Meteorology*, Vol. 131, No. 1, 85-103, ISSN 0006-8314

Lautner, P. (2009) Hungarian Air Quality Network, http://edktvf.zoldhatosag.hu/tartalom/meroall/olm_2009.pdf (In Hungarian)

Lovas, R.; Patvarczki, J.; Kacsuk, P.; Lagzi, I.; Turányi, T; Kullmann, L.; Haszpra, L.; Mészáros, R.; Horányi, A.; Bencsura, A. & Lendvay, Gy. (2006) Air Pollution Forecast on the HUNGRID Infrastructure. In: *Parallel Computing: Current & Future Issues of High-End Computing, Proceedings of the International Conference ParCo 2005*, Joubert, G.R.; Nagel, W.E.; Peters, F.J.; Plata, O.; Tirado, P. & Zapata E. (Eds.), 121-128, ISBN 3-00-017352-8, John von Neumann Institute for Computing, Jülich, NIC Series, Vol. 33

Luhar, A.K. & Patil, R.S. (1989) A general finite line source model for vehicular pollution prediction. *Atmospheric Environment*, Vol. 23, 555-562, ISSN: 1352-2310

Makra, L. (2005) The role of traffic in modifying air quality in a medium-sized city, Szeged, Hungary. *Epidemiology*, Vol. 16, No. 5, S62, ISSN: 1044-3983

Makra, L. & Horváth, Sz. (2001) Assessment of air pollution in Szeged. *Légkör,* 46, 14-18. (in Hungarian), ISSN: 0133-3666

Makra, L.; Horváth, Sz.; Zempléni, A.; Csiszár, V.; Rózsa, K. & Motika, G. (2001a) Air Quality Trends in southern Hungary. *EURASAP Newsletter*, Vol. 42, 2-13, ISSN 1687-1421

Makra, L.; Horváth, Sz.; Zempléni, A.; Csiszár, V.; Fodré, Zs.; Bucsiné Kapocsi, I.; Motika, G. & Sümeghy, Z. (2001b) Analysis of air quality parameters in Csongrád county. *Acta Climatologica Chrologica Universitatis Szegediensis*, Vol. 34-35, 23-44, ISSN: 1587-5903, ISSN: 1587-5903

Makra, L.; Horváth, Sz. & Sümeghy, Z. (2002) An objective analysis and ranking of cities on environmental and social factors. *IGU 2002. Geographical Renaissance at the Dawn of the Millennium.* Durban, South-Africa, 2002. In: Climates in Transition (Ed: Nkemdirim, L.C.), Minuteman Press, 161-172

Makra, L.; Mayer, H.; Bérczi, R. & Borsos, E. (2003) Evaluation of the air quality of Szeged with some assessment methods. *Acta Climatologica et Chrologica Universitatis Szegediensis,* Vol. 36–37, 85-93, ISSN: 1587-5903

Makra, L.; Juhász, M.; Borsos, E. & Béczi, R. (2004) Meteorological variables connected with airborne ragweed pollen in Southern Hungary. *International Journal of Biometeorology,* Vol. 49, 37-47, ISSN: 0020-7128

Makra, L.; Juhász, M.; Béczi, R. & Borsos, E. (2005) The history and impacts of airborne Ambrosia (Asteraceae) pollen in Hungary. *Grana,* Vol. 43, 1-8, ISSN 0017-3134

Makra, L.; Mika, J.; Bartzokas, A.; Béczi, R.; Borsos, E. & Sümeghy, Z. (2006a) An objective classification system of air mass types for Szeged, Hungary with special interest to air pollution levels. *Meteorology and Atmospheric Physics,* Vol. 92, No. 1-2, 115-137, ISSN: 0177-7971

Makra, L.; Juhász, M.; Mika, J.; Bartzokas, A.; Béczi, R. & Sümeghy, Z. (2006b) An objective classification system of air mass types for Szeged, Hungary with special attention to plant pollen levels. *International Journal of Biometeorology,* Vol. 50, No. 6, 403-421, ISSN: 0020-7128

Makra, L.; Mika, J.; Bartzokas, A. & Sümeghy, Z. (2007a) Relationship between the Péczely's large-scale weather types and air pollution levels in Szeged, Southern Hungary. *Fresenius Environmental Bulletin,* Vol. 16, No. 6, 660-673, ISSN: 10184619

Makra, L.; Juhász, M.; Mika, J.; Bartzokas, A.; Béczi, R. & Sümeghy, Z. (2007b) Relationship between the Péczely's large-scale weather types and airborne pollen grain concentrations for Szeged, Hungary. *Grana,* Vol. 46, No. 1, 43-56, ISSN 0017-3134

Makra, L.; Sánta, T. & Baranka, G. (2008) Modeling Air Pollution of Vehicular Traffic in Szeged, Southern Hungary. *Epidemiology,* Vol. 19, No. 6, S85-S86. Suppl. S, ISSN: 1044-3983

Makra, L.; Mika, J.; Bartzokas, A.; Béczi, R. & Sümeghy, Z. (2009) Comparison of objective air-mass types and the Péczely weather types and their ability to classify levels of air pollutants in Szeged, Hungary. *International Journal of Environment and Pollution, "Air Pollution"* Special Issue (Eds: Makra, L. & Kambezidis, H.D.), Vol. 36, No. 1-3, 81-98, ISSN: 0957-4352

Makra, L.; Mayer, H.; Mika, J.; Sánta, T. & Holst, J. (2010) Variations of traffic related air pollution on different time scales in Szeged, Hungary and Freiburg, Germany. *Physics and Chemistry of the Earth,* Vol. 35, No. 1-2, 85-94, ISSN: 1474-7065

Mayer, H. (1999) Air pollution in cities. *Atmospheric Environment,* Vol. 33, 4029-4037 ISSN: 1352-2310

Mayer, H.; Kalberlah, F. & Ahrens, D. (2002a) TLQ - Am impact related air quality index obtained on a daily basis. *Procedings of the Fourth Symposium on the Urban Environment,* 80-81. Norfolk, VA, USA

Mayer, H.; Kalberlah, F.; Ahrens, D. & Reuter, U. (2002b) Analysis of indices for the assessment of the air. *Gefahrstoffe-Reinhaltung der Luft,* Vol. 62, 177-183, ISSN: 0949-8036 (In German)

Mayer, H.; Makra, L.; Kalberlach, F.; Ahrens, D. & Reuter, U. (2004) Air stress and air quality indices. *Meteorologische Zeitschrift,* Vol. 13, 395-403, ISSN: 0941-2948

Mensink, C.; De Ridder, K.; Deutsch, F.; Lefebre, F. & Van de Vel, K. (2008) Examples of scale interactions in local, urban, and regional air quality modeling. *Atmospheric Research,* Vol. 89, 351–357, ISSN: 0169-8095

Mészáros, R.; Vincze, Cs. & Lagzi I. (2010) Simulation of accidental release using a coupled transport (TREX) and numerical weather prediction (ALADIN) model. *Időjárás,* Vol. 114, No. 1-2, 101-120, ISSN: 0324-6329

Ministry of Environment and Water, Hungary (2009) *Data on environment of Hungary.* Budapest, Hungary (Web page of the ministry: http://www.ktm.hu/), ISBN: 963-00-2513-2

Mohl, M.; Gaskó, B.; Horváth, Sz.; Makra, L. & Szabó, F. (2002) *2nd Environmental Programme of Szeged, 2003-2007.* Manuscript (in Hungarian) (Mayor's Office, H-6720 Szeged, Széchenyi tér 10, Hungary)

Mórik, J. (1970) Nearly a century of research on air hygiene in Hungary. *Időjárás,* Vol. 74, No. 5-6, 368-376, ISSN: 0324-6329 (In Hungarian)

Moussiopoulos, N.; Borrego, C.; Bozó, L.; Galmarini, S.; Poppe, D.; Schatzmann, M. & Sturm, P. (2003) Urban and Local Scale Air Pollution. *Towards Cleaner Air for Europe –*

*Science, Tools and Applications*, Midgley, P. & Reuther, M (Eds.), 123-156, ISBN: 3-8236-1390-1, Weikersheim: Margraf Verlag

Möller, D. (1999) Global Problems of Atmospheric Chemistry - The Story of Man's Impact on Atmospheric Ozone. *Atmospheric environmental research: critical decisions between technological Process and Preservation of Nature*. Möller, D. (Ed.), 3-30, ISBN: 3-540-6355-9, Springer-Verlag Berlin, Heidelberg

Müller, G.; Artz, R.; Baltensperger, U.; Carmichael, G.; Dlugokencky, E.; Penkett, S.; Stähelin, J. & Webb, A. (2007) WMO Global Atmosphere Watch (GAW) Strategic Plan: 2008-2015, *GAW Report 172*, WMO TD No. 1384, Geneva, Switzerland

Oetll, D.; Almbauer, R.A. & Sturm, P.J. (2001a) A new method to estimate diffusion in low wind, stable conditions. *Journal of Applied Meteteorology*, Vol. 40, 259-268, ISSN: 0894-8763

Oettl, D.; Kukkonen, J.; Almbauer, R.A.; Sturm, P.J.; Pohjola, M. & Härkönen, J. (2001b) Evaluation of a Gaussian and a Lagrangian model against a roadside dataset, with focus on low wind speed conditions. *Atmospheric Environment*, Vol. 35, 2123-2132, ISSN: 1352-2310

Pavelin, E.G.; Johnson, C.E.; Rughooputh, S. & Toumi, R. (1999) Evaluation of pre-industrial surface ozone measurements made using Schönbein's method. *Atmospheric Environment*, Vol. 33, 919-929, ISSN: 1352-2310

Petersen, W. (1980) 'User's guide for HIWAY2, a highway air pollution model. *EPA-600/8-80-018*, US Environmental Protection Agency, Research Triangle Park, North Carolina.

Péczely, G. (1957) *Grosswetterlagen in Ungarn*. Kleinere Veröffentlichungen der Zentralanstalt für Meteorologie, Budapest, Vol. 30, 86 p. (in German)

Sindosi, O.A.; Katsoulis, B.D. & Bartzokas, A. (2003) An objective definition of air mass types affecting Athens, Greece; the corresponding atmospheric pressure patterns and air pollution levels. *Environmental Technology*, Vol. 24, 947-962.

Steib, R. (2005) Regulatory modelling activity in Hungary, In: *Advances in Air Pollution Modelling for Environmental Security*, Faragó, I.; Georgiev, K. & Havasi, Á. (Ed.), 337-347, Springer, ISBN 10 1-4020-3349-4, Netherlands

Steib, R. & Labancz, K. (2005) Regulatory modeling in Hungary - the AERMOD model. Part I. Description and application. *Időjárás*, Vol. 109, No. 3, 157-172, ISSN: 0324-6329

Steib, R.; Labancz, K.; Ferenczi, Z. & Alföldy, B. (2008) Airport (Budapest Ferihegy - Hungary) air quality analysis using the EDMS modeling system. Part I. Model development and testing. *Időjárás*, Vol. 112, No. 2, 99-112, ISSN: 0324-6329

Strategy for EMEP 2000-2009. *United Nations Publications*. Sales No. 01.II.E.18, Geneva Switzerland ISBN 92-1-116784-1

Szepesi, D.; Fekete, K.; Büki, R.; Koncsos, L. & Kovács, E. (2005) Development of regulatory transmission modeling in Hungary. *Időjárás*, Vol. 109, No. 4, 257-279, ISSN: 0324-6329

Turner, D. B. (1964) A Diffusion Model for an Urban Area. *Journal of Applied Meteorology*, Vol. 3, 83-91, ISSN: 0894-8763

Unger, J. & Pongrácz, R. (2008) Urban Climate Research in Hungary. *Urban Climate News, Quarterly Newsletter of the IAUC*, Vol. 28, 17-20

Warnek, P. (1999) Chemistry of the natural atmosphere. *Academic Press*, p. 927, ISBN 0-12-73-56-32-0, San Diego

Weidinger, T.; Gyöngyösi, A.Z. & Sass, E. (2006) Sensitivity studies of the coupled AERMOD/ETA atmospheric dispersion modeling system. In: *Proceedings of Conference on Modelling Fluid Flow (CMFF'06)*, Lajos, T. and Vad, J. (Eds.), 133-139, Budapest University of Technology and Economics, Budapest

Weidinger, T.; Pinto, J.; Baranka, Gy. & Ivády, A. (2009) Historical ozone Measurements made in the Habsburg Empire during the 19 thcentury. In: *Book of Abstract, XXIII International Congress of History of Science and Technology, Ideas and Instruments in Social Context*, 554 p, Budapest, Hungary

Weidinger T.; Horváth, L.; Nagy Z. & Gyöngyösi A.Z. (2010) Long-term measurements of energy budget and trace gas fluxes between the atmosphere and different types of ecosystems in Hungary. Advances in environmental fluid mechanics (Eds.: Mihailovic, D.T. & Gualtieri, C.) Publisher: *World Scientific*, 185-208, ISBN: 978-981-4291-99-6, New Jersey, London, Singapore, Beijing, Shanghai, Hong Kong, Taipei, Chennai

# 6

# Optimization of Traffic Behavior via Fluid Dynamic Approach

Ciro D'Apice[1], Rosanna Manzo[1] and Benedetto Piccoli[2]
[1]*Department of Information Engineering and Applied Mathematics, University of Salerno, Fisciano (SA)*
[2]*Istituto per le Applicazioni del Calcolo "Mauro Picone", Consiglio Nazionale delle Ricerche, Roma*
*Italy*

## 1. Introduction

The exponentially increasing number of circulating cars in modern cities renders the problem of traffic control of paramount importance. Traffic congestion is a condition on networks that occurs in the presence of an excess of vehicles on a portion of roadway, and is characterized by slower speeds, longer trip times, and increased queueing. As demand approaches the capacity of a road (or of the intersections along the road), extreme traffic congestion sets in. Incidents may cause ripple effects (a cascading failure) which then spread out and create a sustained traffic jam. The presence of hard congestions on urban networks may have dramatic implications, affecting productivity, pollution, life style, the passage of emergency vehicles traveling to their destinations where they are urgently needed. Transportation engineers and emergency planners work together to alleviate congestion and, in addition to traditional efforts, an increased focus is addressed to the development and promotion of transportation systems management and operations. The main inspiration is the understanding and optimization of traffic behavior in order to answer to several questions: where to install traffic lights or stop signs; how long the cycle of traffic lights should be; how to distribute flows at junctions, where to construct entrances, exits and overpasses, etc. in order to maximize cars flow, minimize traffic congestions, accidents, pollution.

The problem of modeling car traffic has been faced resorting to different approaches ranging from microscopic ones, taking into account each single car, to kinetic and macroscopic fluid-dynamic ones, dealing with traffic situations resulting from the complex interaction of many vehicles. Each of them implies some technical approximations, and suffers therefore from related drawbacks, either analytical or computational. Here we are interested to traffic flow on a road network, modelled by a fluid-dynamic approach.

In the 1950s James Lighthill and Gerard Whitham, two experts in fluidynamics, and independently P. Richards, modeled the flow of car traffic along a single road using the same equations describing the flow of water (Lighthill et al. (1955); Richards (1956)). The basic idea is to look at large scales so as to consider cars as small particles and to assume the conservation of the cars number. The LWR model is described by a single conservation law, a special partial differential equation where the variable, the car density, is a conserved quantity, i.e. a quantity which can neither be created or destroyed. Then some second order models, i.e. with two equations, were proposed by Payne and Whitham (Payne (1971; 1979); Whitham (1974)). Since the assumption of the LWR model of the dependence of the average speed $v$ only on the density $\rho$ is not valid in some situations, Payne and Whitham introduced an additional equation for the speed, including a relaxation term for $v$. Unfortunately this model suffers from sever drawbacks, which led Daganzo in 1995 to write a celebrated "requiem" for this kind of second order approximation of traffic flow (Daganzo (1995)). In particular he proved that cars may exhibit negative speed and the model violates the so-called anisotropy principle, i.e., the fact that a car should be influenced only by the traffic dynamics ahead of it, being practically insensitive to what happens behind. Finally Aw and Rascle in 2000, to overcome Daganzo's observations, proposed a "resurrection" of second order models, introducing an equation for the pressure as increasing function of the density (Aw & Rascle (2000)). The Aw Rascle model gave origin to a lot of other traffic models and derivations. The first third order model was proposed by Helbing (see Helbing (2001)). Colombo in 2002 developed an hyperbolic phase transition model, in which the existence of the phase transition is postulated and accounted for by splitting the state space $(\rho, f)$, where $f$ is the flux, in two regions, corresponding to the regimes of free and congested flow (Colombo (2002 a;b)). A multilane extension of the Aw-Rascle model was proposed by Greenberg, Klar and Rascle (see Greenberg et al. (2003)). The idea to consider the LWR model on a network was proposed by Holden and Risebro (Holden et al. (1995)). They solved the Riemann Problem at junctions (the problem with constant initial data on each road), proposing a maximization of the flux. Existence of solution to Cauchy Problems and the counterexample to the Lipschitz continuous dependence on initial data was proved in the paper by Coclite et al. (2005). The Aw-Rascle second order model has been extended to networks in Garavello & Piccoli (2006 b).

Traffic congestion leads to a strong degradation of the network infrastructure and accordingly reduced throughput, which can be countered via suitable control measures and strategies. Some optimization problems for road networks modeled by fluid-dynamic approach have been already studied: Helbing et al. (2005) is devoted to traffic light regulation, while Gugat et al. (2005) and Herty et al. (2003) are more related to our analysis but focus on the case of smooth solutions (not developing shocks) and boundary control. A specific traffic regulation problem is addressed in Chitour & Piccoli (2005). Given a crossing with some expected traffic, is it preferable to construct a traffic circle or a light? The two solutions are studied in terms of flow control and the performances are compared.

In this Chapter we report some recent optimization results obtained in Cascone et al. (2007; 2008 a;b); Cutolo et al. (2009); Manzo et al. (2010) for urban traffic networks, whose evolution is described by the LWR model.

Road networks consist of a finite set of roads, that meet at some junctions. The dynamics is governed on each road by a conservation law. In order to uniquely solve the Riemann Problem at junctions and to construct solutions via Wave Front Tracking (see Bressan (2000); Garavello & Piccoli (2006 a)), as the system is under-determined even after imposing the conservation of cars, the following assumptions are made: the incoming traffic distributes to outgoing roads according to fixed (statistical) distribution coefficients; drivers behave in order to maximize the through flux. More precisely, if the number of incoming roads is greater than that of outgoing roads, one has also to introduce right of way parameters.

Some cost functionals have been defined to analyze the traffic behavior: $J_1$ measuring car average velocity, $J_2$ the average traveling time, $J_3$ the total flux of cars, $J_4$ the car density, $J_5$, the Stop and Go Waves functional (*SGW*), the velocity variation, $J_6$ the kinetic energy, and finally $J_7$ measuring the average traveling time weighted with the number of cars moving on each road. Notice that the cost functionals are evaluated through the use of a linear decreasing velocity function $v(\rho) = 1 - \rho$.
For a fixed time horizon $[0,T]$, $\int_0^T J_1(t)\,dt$, $\int_0^T J_3(t)\,dt$, $\int_0^T J_6(t)\,dt$ have been maximized and $\int_0^T J_2(t)\,dt$, $\int_0^T J_7(t)\,dt$ minimized, choosing as controls the right of way parameters or the distribution coefficients depending on the junctions type. A junction of $n \times m$ type is a junction with $n$ incoming roads and $m$ outgoing ones. The attention has been focused on a decentralized approach reducing the analysis of a network to simple junctions. We computed the optimal parameters for single junctions of type $1 \times 2$ and $2 \times 1$ and every initial data. For a complex network, we used the (locally) optimal parameters at every junction and we verified the performance of the (locally) optimal parameters comparing, via simulations, with other choices as fixed and random parameters. The optimization problem for junctions of $2 \times 1$ type, using as control the right of way parameter $p$, every initial data and the functionals $J_i$, $i = 1, 2, 6, 7$ (while $J_3$ happens to be constant) is solved in Cascone et al. (2007) and Cutolo et al. (2009). In particular the functionals $J_2$ and $J_7$ are maximized for the same values of $p$, while $J_1$ and $J_6$ have, in some cases, different optimal values. It is interesting to notice that in many cases there is a set of optimal values of the right of way parameters. Optimization results have been achieved for the functionals $J_i$, $i = 1, 2,3$ (see Cascone et al. (2008 a)) and $J_6$, $J_7$ in the case of junctions of type $1 \times 2$, using the distribution coefficient as control. Observe that the functionals $J_6$ and $J_7$ are optimized for the same values of the distribution coefficient which maximize and minimize $J_1$ and $J_2$, respectively. All the results have been tested by simulations on case studies.
Recently the problem of traffic redirection in the case an accident occurs in a congested area has been considered, see Manzo et al. (2010). Fire, police, ambulance, repair crews, emergency and life-saving equipment, services and supplies must move quickly to where the greatest need is. Assuming that emergency vehicles will cross a given incoming road $I_\varphi$, $\varphi \in \{1,2\}$ and a given outgoing road $I_\psi$, $\psi \in \{3,4\}$ of a junction of type $2 \times 2$, a cost functional measuring the average velocities of such vehicles on the assigned path is analyzed. The optimization results give the values of $\alpha$ and $\beta$ (respectively, the probability that drivers go from road 1 to road 3 and from road 2 to road 3) which maximize the functional, allowing a fast transit of emergency vehicles to reach car accidents places and hospitals.
The Chapter is organized as follows. Section 2 reports the model for road networks. Riemann Solvers at junctions are described in Section 3. The subsequent Section 4 is devoted to the definition of the functionals, introduced to measure network performance. In particular in the Subsection 4.1 we optimize right of way parameters for 2×1 junctions, while in Subsection 4.1 we report optimization studies for $1 \times 2$ junctions. The Section 5 deals with some new results on the optimal redistribution of flows at nodes of type 2×2 in order to maximize the velocity of emergency vehicles on assigned paths. In all the Sections simulations are presented and discussed to illustrate the analytical optimization results.

## 2. Mathematical model for road networks

We consider a network, that is modelled by a finite set of roads $I_k = [a_k, b_k] \subset \mathbb{R}$, $k = 1, \ldots, N$, $a_k < b_k$, possibly with either $a_k = -\infty$ or $b_k = +\infty$. We assume that roads are connected at

junctions. Each junction $J$ is characterized by a finite number $n$ of incoming roads and a finite number $m$ of outgoing ones, thus we identify $J$ with $((i_1, \ldots, i_n), (j_1, \ldots, j_m))$. Hence, the complete model is given by a couple $(\mathcal{I},\mathcal{J})$, where $\mathcal{I} = \{I_k : k = 1, \ldots,N\}$ is the collection of roads and $\mathcal{J}$ is the collection of junctions. The main dependent variables introduced to describe mathematically the problem are the density of cars $\rho = \rho(t, x)$ and their average velocity $v = v(t, x)$ at time $t$ in the point $x$. From these quantities another important variable is derived, namely the flux $f = f(t, x)$ given by $f = \rho v$, which is of great interest for both theoretical and experimental purposes.
On each single road, the evolution is governed by the scalar conservation law:

$$\partial_t \rho + \partial_x f(\rho) = 0, \tag{1}$$

where $\rho = \rho(t, x) \in [0,\rho_{\max}]$, $(t, x) \in \mathbb{R}^2$, with $\rho_{\max}$ the maximal density of cars.
The network load is described by a finite set of functions $\rho_k$ defined on $[0,+\infty[ \times I_k$. On each road $I_k$ we require $\rho_k$ to be a weak entropic solution of the conservation law (1), that is such that for every smooth, positive function $\varphi$: $[0,+\infty[ \times I_k \rightarrow \mathbb{R}$ with compact support on $]0,+\infty[ \times ]a_k, b_k[$

$$\int_0^{+\infty}\int_{a_k}^{b_k}\left(\rho_k \frac{\partial \varphi}{\partial t} + f(\rho_k)\frac{\partial \varphi}{\partial x}\right)dxdt = 0, \tag{2}$$

and entropy conditions are verified, see Bressan (2000); Dafermos (1999); Serre (1996).
It is well known that, for equation (1) on $\mathbb{R}$ and for every sufficiently small initial data in $BV$ (here $BV$ stands for bounded variation functions), there exists a unique weak entropic solution depending in a continuous way from the initial data in $L^1_{loc}$. Moreover, for initial data in $L^\infty \cap L^1$ Lipschitz continuous dependence in $L^1$ is achieved.
Now we discuss how to define solutions at junctions. For this, fix a junction $J$ with $n$ incoming roads, say $I_1, \ldots, I_n$, and $m$ outgoing ones, say $I_{n+1}, \ldots, I_{n+m}$ (briefly a junction of type $n \times m$). A weak solution at $J$ is a collection of functions $\rho_l$: $[0,+\infty[ \times I_l \rightarrow \mathbb{R}$, $l = 1, \ldots,n + m$, such that

$$\sum_{l=0}^{n+m}\left(\int_0^{+\infty}\int_{a_l}^{b_l}\left(\rho_l \frac{\partial \varphi_l}{\partial t} + f(\rho_l)\frac{\partial \varphi_l}{\partial x}\right)dxdt\right) = 0, \tag{3}$$

for every smooth $\varphi_l$, $l = 1, \ldots,n + m$, having compact support in $]0,+\infty[ \times ]a_l, b_l]$ for $l = 1, \ldots,n$ and in $]0,+\infty[ \times [a_l, b_l[$ for $l = n + 1, \ldots,n + m$, also smooth across the junction, i.e.,

$$\varphi_i(\cdot,b_i) = \varphi_j(\cdot,a_j), \quad \frac{\partial \varphi_i}{\partial x}(\cdot,b_i) = \frac{\partial \varphi_j}{\partial x}(\cdot,a_j), \quad i = 1,\ldots,n, j = n+1,\ldots,n+m.$$

A weak solution $\rho$ at $J$ satisfies the Rankine-Hugoniot condition at the junction, namely

$$\sum_{i=1}^{n} f(\rho_i(t,b_i-)) = \sum_{j=n+1}^{n+m} f(\rho_j(t,a_j+)),$$

for almost every $t > 0$. This Kirchhoff type condition ensures the conservation of $\rho$ at junctions. For a system of conservation laws on the real line, a Riemann problem (RP) is a

Cauchy Problem (CP) for an initial datum of Heavyside type, that is piecewise constant with only one discontinuity. One looks for centered solutions, i.e. $\rho(t, x) = \phi(\frac{x}{t})$ formed by simple waves, which are the building blocks to construct solutions to the CP via Wave Front Tracking (WFT) algorithms. These solutions are formed by continuous waves called rarefactions and by traveling discontinuities called shocks. The speeds of the waves are related to the eigenvalues of the Jacobian matrix of $f$, see Bressan (2000). Analogously, we call RP for a junction the Cauchy Problem corresponding to initial data which are constant on each road. The discontinuity in this case is represented by the junction itself.

**Definition 1** *A Riemann Solver (RS) for the junction J is a map* $RS : \mathbb{R}^n \times \mathbb{R}^m \rightarrow \mathbb{R}^n \times \mathbb{R}^m$ *that associates to Riemann data* $\rho_0 = (\rho_{1,0}, \ldots, \rho_{n+m,0})$ *at J a vector* $\hat{\rho} = (\hat{\rho}_1, \ldots, \hat{\rho}_{n+m})$*, so that the solution on an incoming road* $I_i$, $i = 1, \ldots, n$, *is given by the waves produced by the RP* $(\rho_i, \hat{\rho}_i)$, *and on an outgoing road* $I_j$, $j = n + 1, \ldots, n + m$, *by the waves produced by the RP* $(\hat{\rho}_j, \rho_j)$. *We require the consistency condition*

*(CC)* $RS(RS(\rho_0)) = RS(\rho_0)$.

A *RS* is further required to guarantee the fulfillment of the following properties:

(H1) The waves generated from the junction must have negative velocities on incoming roads and positive velocities on outgoing ones.

(H2) Relation (3) holds for solutions to RPs at the junction.

(H3) The map $\rho_0 \mapsto f(\hat{\rho})$ is continuous.

Condition (H1) is a consistency condition to well describe the dynamics at junction. In fact, if (H1) does not hold, then some waves generated by the *RS disappear* inside the junction. Condition (H2) is necessary to have a weak solution at the junction. However, in some cases (H2) is violated if only some components of $\rho$ have to be conserved at the junction, see for instance Garavello & Piccoli (2006 b). Finally, (H3) is a regularity condition, necessary to have a well-posed theory. The continuity of the map $\rho_0 \mapsto f(\hat{\rho})$ can not hold in case (H1) holds true.

There are some important consequences of property (H1), in particular some restrictions on the possible values of fluxes and densities arise. Consider, for instance, a single conservation law for a bounded quantity, e.g. $\rho \in [0, \rho_{max}]$, and assume the following:

**(F)** The flux function $f : [0, \rho_{max}] \mapsto \mathbb{R}$ is strictly concave, $f(0) = f(\rho_{max}) = 0$, thus $f$ has a unique maximum point $\sigma$.

Fixing $\rho_{max} = 1$, one example of velocity function whose corresponding flux ensures (F) is:

$$v(\rho) = 1 - \rho. \tag{4}$$

Then the flux is given by

$$f(\rho) = \rho(1 - \rho). \tag{5}$$

Defining:

**Definition 2** *Let* $\tau : [0, \rho_{max}] \rightarrow [0, \rho_{max}]$ *be the map such that* $f(\tau(\rho)) = f(\rho)$ *for every* $\rho \in [0, \rho_{max}]$, *and* $\tau(\rho) \neq \rho$ *for every* $\rho \in [0, \rho_{max}] \setminus \{\sigma\}$,

we get the following:

**Proposition 3** *Consider a single conservation law for a bounded quantity* $\rho \in [0, \rho_{max}]$ *and assume (F). Let RS be a Riemann Solver for a junction,* $\rho_0 = (\rho_{i,0}, \rho_{j,0})$ *the initial datum and* $RS(\rho_0) = \hat{\rho} = (\hat{\rho}_i, \hat{\rho}_j)$. *Then,*

$$\hat{\rho}_i \in \begin{cases} \{\rho_{i,0}\} \cup \left]\tau(\rho_{i,0}),\rho_{max}\right], & if \quad 0 \le \rho_{i,0} \le \sigma, \\ \left[\sigma,\rho_{max}\right], & if \quad \sigma \le \rho_{i,0} \le \rho_{max}, \end{cases} \quad i = 1,...,n,$$

$$\hat{\rho}_j \in \begin{cases} [0,\sigma], & if \quad 0 \le \rho_{j,0} \le \sigma, \\ \{\rho_{j,0}\} \cup [0,\tau(\rho_{i,0})[, & if \quad \sigma \le \rho_{j,0} \le \rho_{max}, \end{cases} \quad j = n+1,...,n+m.$$

Thanks to Proposition 3, we have the following:

**Proposition 4** *Consider a single conservation law for a bounded quantity* $\rho \in [0,\rho_{max}]$ *and assume (F). To define a RS at a junction J, fulfilling rule (H1), it is enough to assign the flux values* $f(\hat{\rho})$. *Moreover, there exist maximal possible fluxes given by:*

$$f_i^{max}(\rho_0) = \begin{cases} f(\rho_{i,0}), & if \quad 0 \le \rho_{i,0} \le \sigma, \\ f(\sigma), & if \quad \sigma \le \rho_{i,0} \le \rho_{max}, \end{cases} \quad i = 1,...,n,$$

$$f_j^{max}(\rho_0) = \begin{cases} f(\sigma), & if \quad 0 \le \rho_{j,0} \le \sigma, \\ f(\rho_{j,0}), & if \quad \sigma \le \rho_{j,0} \le \rho_{max}, \end{cases} \quad j = n+1,...,n+m.$$

Once a Riemann Solver $RS_J$ at a junction $J$ is assigned, we define admissible solutions at $J$ those $\rho$ such that $t \mapsto \rho(t, \cdot)$ is $BV$ for almost every $t$, and moreover:

$$RS(\rho_J(t)) = \rho_J(t),$$

where

$$\rho_J(t) = (\rho_1(\cdot, b_1-), \ldots, \rho_n(\cdot, b_n-), \rho_{n+1}(\cdot, a_{n+1}+), \ldots, \rho_{n+m}(\cdot, a_{n+m}+)).$$

For every road $I_k = [a_k, b_k]$, such that either $a_k > -\infty$ and $I_k$ is not the outgoing road of any junction, or $b_k < +\infty$ and $I_k$ is not the incoming road of any junction, a boundary datum $\psi_k$ : $[0,+\infty[\to\mathbb{R}^n$ is given. We require $\rho_k$ to satisfy $\rho_k(t, a_k) = \psi_k(t)$ (or $\rho_k(t, b_k) = \psi_k(t)$) in the sense of Bardos et al. (1979). For simplicity, we assume that boundary data are not necessary. The aim is to solve the CP for a given initial datum as in the next definition.

**Definition 5** *Given* $\bar{\rho}_k : I_k \to\mathbb{R}^n$, $k = 1, ...,N$, *in* $L^1_{loc}$, *a collection of functions* $\rho = (\rho_1, ...,\rho_N)$, *with* $\rho_k : [0,+\infty[\times I_k \to\mathbb{R}^n$ *continuous as function from* $[0,+\infty[$ *into* $L^1_{loc}$, *is an admissible solution to the Cauchy Problem on the network if* $\rho_k$ *is a weak entropic solution to (1) on* $I_k$, $\rho_k(0, x) = \bar{\rho}_k(x)$ *a.e. and at each junction* $\rho$ *is an admissible solution.*

There is a general strategy, based on Wave Front Tracking, to prove existence of solution on a whole network for CPs.

The main steps are the following (see Garavello & Piccoli (2006 a;b) for details):

1. Construct approximate solutions via WFT algorithms, using the *RS* at junctions for interaction of waves with junctions.
2. Estimate the variation of flux for interaction of waves with junctions, thus on the whole network.
3. Pass to the limit using the previous steps.

In what follows we suppose that $f_k = f$, $\forall k = 1, ...,N$, but it is possible to generalize all definitions and results to the case of different fluxes $f_k$ for each road $I_k$. In fact, all statements are in terms of values of fluxes at junctions.

## 3. Riemann Solvers according to rule (RA)

We assume that (F) holds true and we look for Riemman Solvers fulfilling (H1). Thus, in view of Proposition 4, it is enough to determine the fluxes values.
In Coclite et al. (2005) an RS at junctions is considered, based on the following algorithm:
(RA) We assume that
(A) the traffic from incoming roads is distributed on outgoing ones according to fixed coefficients;
(B) fulfilling (A), the through flux is maximized.
Consider a junction of $n\times m$ type. For simplicity we use the notation $\gamma_k = f(\rho_k)$, $\gamma_k^{\max} = f_k^{\max}$, $\hat{\gamma}_k = f(\hat{\rho}_k)$, $k = 1, \ldots, n + m$.
If the incoming roads are $I_1, \ldots, I_n$ and the outgoing ones $I_{n+1}, \ldots, I_{n+m}$, rule (A) corresponds to fix a stochastic matrix $A = (\alpha_{j,i})$ where $j = n + 1, \ldots, n + m$ and $i = 1, \ldots, n$. The coefficient $\alpha_{j,i}$ represents the percentage of traffic from $I_i$ directed to $I_j$. Here we assume:

$$0 < \alpha_{j,i} < 1, \quad \sum_{j=1}^{m} \alpha_{j,i} = 1.$$

Recalling Proposition 4, we define:

$$\Omega_{in} = [0, \gamma_1^{max}] \times \cdots \times [0, \gamma_n^{max}], \qquad \Omega_{out} = [0, \gamma_{n+1}^{max}] \times \cdots \times [0, \gamma_{n+m}^{max}].$$

From (H1), one gets that the incoming fluxes must take values in $\Omega_{in}$ and the outgoing fluxes in $\Omega_{out}$. Moreover, in order to fulfill rule (A), the incoming fluxes must belong to the region:

$$\tilde{\Omega}_{in} = \{\gamma \in \Omega_{in} : A \cdot \gamma \in \Omega_{out}\}.$$

Notice that $\tilde{\Omega}_{in}$ is a convex set determined by linear constraints. Moreover, rule (A) implies (H2). Thus rule (B) is equivalent to maximize only over incoming fluxes, then outgoing ones can be determined by rule (A). Finally, rules (A) and (B) correspond to a Linear Programming problem: Maximize the sum of fluxes from incoming roads over the region $\tilde{\Omega}_{in}$. Such problem always admits a solution, which is unique provided the cost function gradient (here the vector with all components equal to 1) is not orthogonal to the linear constraints describing the set $\tilde{\Omega}_{in}$.
Let us now consider a junction of type 1 ×2. In detail, 1 is the only incoming road, while 2 and 3 are the outgoing roads. The distribution matrix $A$ takes the form

$$A = \begin{pmatrix} \alpha \\ 1-\alpha \end{pmatrix},$$

where $\alpha \in \, ]0,1[$ and $1 - \alpha$ indicate the percentage of cars which, from road 1, goes to road 2 and 3, respectively. Thanks to rule (B), the solution to a RP is:

$$\hat{\gamma} = (\hat{\gamma}_1, \hat{\gamma}_2, \hat{\gamma}_3) = (\hat{\gamma}_1, \alpha\hat{\gamma}_1, (1-\alpha)\hat{\gamma}_1),$$

where

$$\hat{\gamma}_1 = \min\left\{\gamma_1^{\max}, \frac{\gamma_2^{\max}}{\alpha}, \frac{\gamma_3^{\max}}{1-\alpha}\right\}.$$

For a junction of 2 ×2 type, i.e. with two incoming roads, 1 and 2, and two outgoing roads, 3 and 4, the traffic distribution matrix $A$ assumes the form:

$$A = \begin{pmatrix} \alpha & \beta \\ 1-\alpha & 1-\beta \end{pmatrix},$$

where $\alpha$ is the probability that drivers go from road 1 to road 3 and $\beta$ is the probability that drivers travel from road 2 to road 3. Let us suppose that $\alpha \neq \beta$ in order to fulfill the orthogonal condition for uniqueness of solutions. From rule (A), it follows that $\hat{\gamma}_3 = \alpha\hat{\gamma}_1 + \beta\hat{\gamma}_2$, $\hat{\gamma}_4 = (1 - \alpha)\hat{\gamma}_1 + (1 - \beta)\hat{\gamma}_2$. From rule (B), we have that $\hat{\gamma}_\varphi$, $\varphi = 1, 2$, is found solving the Linear Programming problem:

$$\max\ (\gamma_1 + \gamma_2),$$
$$0 \le \gamma_\varphi \le \gamma_\varphi^{\max}, 0 \le \alpha\gamma_1 + \beta\gamma_2 \le \gamma_3^{\max}, 0 \le (1-\alpha)\gamma_1 + (1-\beta)\gamma_2 \le \gamma_4^{\max}.$$

The orthogonality condition can not hold if $n > m$. If not all traffic can flow to the only outgoing road, then one should assign a yielding or priority rule:

**(C)** There exists a priority vector $p \in \mathbb{R}^n$ such that the vector of incoming fluxes must be parallel to $p$.

Let us show how rule (C) works in the simple case $n = 2$ and $m = 1$, i.e. a junction of type 2×1. In this case, the matrix $A$ reduces to the vector (1,1), thus no information is obtained. Rule (B) amounts to determining the through flux as $\Gamma = \min\{\gamma_1^{\max} + \gamma_2^{\max}, \gamma_3^{\max}\}$. If $\Gamma = \gamma_1^{\max} + \gamma_2^{\max}$, then we simply take the maximal flux over both incoming roads. If the opposite happens, consider the space $(\gamma_1, \gamma_2)$ of possible incoming fluxes and define the following lines:

$$r_p : \{tp : t \in \mathbb{R}\},$$

$$r_\Gamma : \{(\gamma_1, \gamma_2) : \gamma_1 + \gamma_2 = \Gamma\}.$$

Let $P$ be the point of intersection of the lines $r_p$ and $r_\Gamma$. Recall that the final fluxes should belong to the region:

$$\Omega = \left\{(\gamma_1, \gamma_2) : 0 \le \gamma_i \le \gamma_i^{\max}, i = 1,2\right\}.$$

We distinguish two cases:

a) $P$ belongs to $\Omega$;

b) $P$ is outside $\Omega$.

In the first case $(\hat{\gamma}_1, \hat{\gamma}_2) = P$, while in the second case $(\hat{\gamma}_1, \hat{\gamma}_2) = Q$, where $Q = proj_{\Omega \cap r_\Gamma}(P)$, and *proj* is the usual projection on a convex set.

The reasoning can be repeated also in the case of $n$ incoming roads. In $\mathbb{R}^n$, the line $r_p$ is again given by $r_p = tp$, $t \in \mathbb{R}$, and

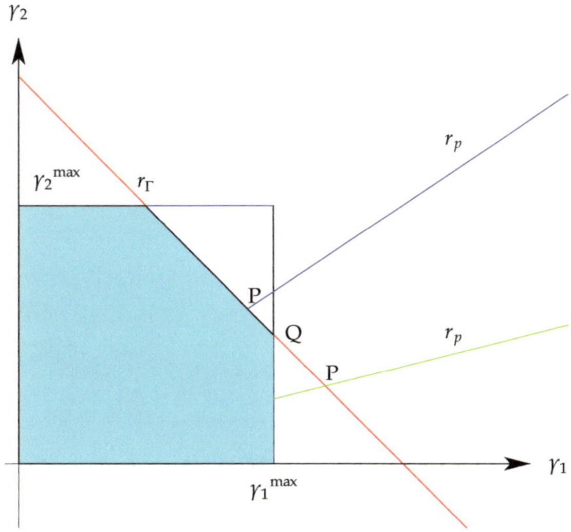


Fig. 1. The two cases: $P$ belongs to $\Omega$ and $P$ is outside $\Omega$.

$$H_\Gamma = \left\{ (\gamma_1, \ldots, \gamma_n) : \sum_{i=1}^{n} \gamma_i = \Gamma \right\}$$

is a hyperplane. There exists a unique point $P = r_p \cap H_\Gamma$. If $P \in \Omega$, then again we use $P$ to determine the incoming fluxes. Otherwise, we choose the point $Q = proj_{\Omega \cap H_\Gamma}(P)$, the projection over the subset $\Omega \cap H_\Gamma$. Notice that the projection is unique since $\Omega \cap H_\Gamma$ is a closed convex subset of $H_\Gamma$. It is easy to check that (H3) is verified for this RS.

## 4. Cost functionals

We focus on a single junction with $n$ incoming roads and $m$ outgoing ones. To evaluate networks performance we define the following functionals:

$J_1$ measuring car average velocity:

$$J_1(t) = \sum_{k=1}^{n+m} \int_{I_k} v(\rho_k(t,x))\,dx,$$

$J_2$ measuring average traveling time:

$$J_2(t) = \sum_{k=1}^{n+m} \int_{I_k} \frac{1}{v(\rho_k(t,x))}\,dx,$$

$J_3$ measuring total flux of cars:

$$J_3(t) = \sum_{k=1}^{n+m} \int_{I_k} f(\rho_k(t,x))\,dx,$$

$J_4$ measuring car density:

$$J_4(t)=\sum_{k=1}^{n+m}\int_0^t\int_{I_k}\rho_k(\tau,x)d\tau dx,$$

$J_5$, the Stop and Go Waves functional, measuring the velocity variation:

$$J_5(t)=SGW=\sum_{k=1}^{n+m}\int_0^t\int_{I_k}|Dv(\rho)|d\tau dx,$$

where $|Dv|$ is the total variation of the distributional derivative $D\rho$,
$J_6$ measuring the kinetic energy:

$$J_6(t)=\sum_{k=1}^{n+m}\int_{I_k}f(\rho_k(t,x))v(\rho_k(t,x))dx,$$

$J_7$ measuring the average traveling time weighted with the number of cars moving on each road $I_k$:

$$J_7(t)=\sum_{k=1}^{n+m}\int_{I_k}\frac{\rho_k(t,x)}{v(\rho_k(t,x))}dx.$$

Given a junction of type $1\times m$ or $n\times 1$, and initial data, solving the RP we determine the average velocity, the average traveling time and the flux over the network as function of the distribution coefficients or the right of way parameters. It follows that also the functionals $J_k$, $k=1, 2, 3, 6,7$ are functions of the same parameters.
For a fixed time horizon $[0,T]$, our aim is to maximize $\int_0^T J_1(t)\,dt$, $\int_0^T J_3(t)\,dt$, $\int_0^T J_6(t)\,dt$ and to minimize $\int_0^T J_2(t)\,dt$, $\int_0^T J_7(t)\,dt$, choosing the right of way parameters $p_k(t)$ or the distribution coefficients $\alpha_k(t)$. Since the solutions of such optimization control problems are too difficult, we reduce to the following problem:

(P) Consider a junction $J$ of $2\times 1$ type or $1\times 2$ type, the functionals $J_k$, $k=1, 2, 3, 6, 7$, and the right of way parameter $p_k(t)$ or the distribution coefficients $\alpha_k(t)$ as controls. We want to minimize $J_2(T)$, $J_7(T)$ and to maximize $J_1(T)$, $J_3(T)$, $J_6(T)$ for $T$ sufficiently big.

As was proved in Cascone et al. (2007; 2008 a;b), the functional $J_3(T)$ does not depend on the right of way parameters and on distribution coefficients.
The optimization approach we followed is of decentralized type. In fact, the optimization is done over $p$ or $\alpha$ for a single junction. For complex networks we adopt the following strategy:

**Step 1.** Compute the optimal parameters for single junctions and every initial data. For this, consider the asymptotic solution over the network (assuming infinite length roads so to avoid boundary data effects).

**Step 2.** Use the (locally) optimal parameters at every junction of the network, updating the value of the parameters at every time instant using the actual density on roads near the junction.

**Step 3.** Verify the performance of the (locally) optimal parameters comparing, via simulations, with other choices as fixed and random parameters.

All the optimization results reported in the following Subsections are obtained assuming the flux function (5).

### 4.1 Optimization of junctions of 2 ×1 type

We focus on junctions of 2 ×1 type, labelling with 1 and 2 the incoming roads and with 3 the outgoing one and we consider $p$ as control (for more details see Cascone et al. (2007; 2008 b); Cutolo et al. (2009)). From the flux function we can express $\hat{\rho}_k$ in terms of $\hat{\gamma}_k$ :

$$\hat{\rho}_k = \frac{1 + s_k\sqrt{1 - 4\hat{\gamma}_k}}{2}, k = 1,2,3,$$

with

$$s_i = \begin{cases} -1, & if \ \ \rho_{i,0} < \sigma, \gamma_1^{\max} + \gamma_2^{\max} \le \gamma_3^{\max}, \\ & or \ \ \rho_{i,0} < \sigma, \gamma_3^{\max} < \gamma_1^{\max} + \gamma_2^{\max}, p_i\hat{\gamma}_3 \ge \gamma_i^{\max}, \\ +1, & if \ \ \rho_{i,0} \ge \sigma, \\ & or \ \ \rho_{i,0} < \sigma, \gamma_3^{\max} < \gamma_1^{\max} + \gamma_2^{\max}, p_i\hat{\gamma}_3 < \gamma_i^{\max}, \end{cases} \quad i = 1,2,$$

$$s_3 = \begin{cases} -1, & if \ \ \rho_{3,0} \le \sigma, \\ & or \ \ \rho_{3,0} > \sigma, \gamma_1^{\max} + \gamma_2^{\max} < \gamma_3^{\max}, \\ +1, & if \ \ \rho_{3,0} > \sigma, \gamma_1^{\max} + \gamma_2^{\max} \ge \gamma_3^{\max}, \end{cases}$$

where

$$p_i = \begin{cases} p, & \text{if } i = 1, \\ 1 - p, & \text{if } i = 2. \end{cases}$$

The velocity, in terms of $\hat{\gamma}_k$, is given by $v(\hat{\rho}_k) = \frac{1 - s_k\sqrt{1 - 4\hat{\gamma}_k}}{2}, k = 1,2,3.$ The functionals $J_2(T)$ and $J_7(T)$ are maximized for the same values of $p$. In fact we get:

**Theorem 6** *Consider a junction J of 2×1 type. For T sufficiently big, the cost functionals $J_2(T)$ and $J_7(T)$ are optimized for the following values of p:*

1. *Case $s_1 = s_2 = +1$, we have that:*
   a. $p = \frac{1}{2}$, *if* $\beta^- \le 1 \le \beta^+$ *or* $\gamma_2^{\max} = \gamma_3^{\max}$;
   b. $p \in [0, p^-]$, *if* $\beta^- \le \beta^+ \le 1$;
   c. $p \in [p^+, 1]$, *if* $1 \le \beta^- \le \beta^+$;
2. *Case $s_1 = -1 = -s_2$, we have that:*
   a. $p = \frac{1}{2}$ *or* $p \in [p^+, 1]$, *if* $\beta^- \le 1 \le \beta^+$ *or* $\gamma_2^{\max} = \gamma_3^{\max}$;
   b. $p \in [0, p^-]$ *or* $p \in [p^+, 1]$, *if* $\beta^- \le \beta^+ \le 1$;
   c. $p \in [p^+, 1]$, *if* $1 \le \beta^- \le \beta^+$;
3. *Case $s_1 = +1 = -s_2$, we have that:*
   a. $p = \frac{1}{2}$ *or* $p \in [0, p^-]$, *if* $\beta^- \le 1 \le \beta^+$;

b. $p \in [0, p^-]$, *if* $\beta^- \le \beta^+ \le 1$;

c. $p \in [0, p^-]$ *or* $p \in [p^+, 1]$, *if* $1 \le \beta^- \le \beta^+$;

d. $p = \frac{1}{2}$ *or* $p \in [p^+, 1]$, *if* $\gamma_2^{max} = \gamma_3^{max}$;

4. *Case* $s_1 = s_2 = -1$, *we have that:*

a. $p = \frac{1}{2}$ *or* $p \in [p^+, 1]$, *if* $\beta^- \le 1 \le \beta^+$, *with* $\beta^- \beta^+ > 1$ *or* $\gamma_2^{max} = \gamma_3^{max}$;

b. $p = \frac{1}{2}$ *or* $p \in [0, p^-]$, *if* $\beta^- \le 1 \le \beta^+$, *with* $\beta^- \beta^+ < 1$;

c. $p = \frac{1}{2}$ *or* $p \in [0, p^-] \cup [p^+, 1]$, *if* $\beta^- \le 1 \le \beta^+$, *with* $\beta^- \beta^+ = 1$;

d. $p \in [0, p^-]$, *if* $\beta^- \le \beta^+ \le 1$;

e. $p \in [p^+, 1]$, *if* $1 \le \beta^- \le \beta^+$;

*where* $\beta^- = \frac{\gamma_3^{max} - \gamma_1^{max}}{\gamma_1^{max}}, \beta^+ = \frac{\gamma_2^{max}}{\gamma_3^{max} - \gamma_2^{max}}, p^+ = \frac{\gamma_1^{max}}{\gamma_3^{max}}, p^- = \frac{\gamma_3^{max} - \gamma_2^{max}}{\gamma_3^{max}}$.

In the particular case $\gamma_1^{max} = \gamma_2^{max} = \gamma_3^{max}$ the functionals $J_2$ and $J_7$ are optimized for $p = \frac{1}{2}$. The maximization of the functionals $J_1(T)$ and $J_6(T)$ is reached, in some cases, for different values of the right of way parameter, as reported in the following theorem, in which the optimization analysis of the new functional $J_6(T)$ is compared with the results obtained in Cascone et al. (2007) for $J_1(T)$.

**Theorem 7** *Consider a junction J of* $2\times1$ *type. For T sufficiently big, the cost functionals* $J_1(T)$ *and* $J_6(T)$ *are optimized for the following values of p:*

1. *Case* $s_1 = s_2 = +1$, *we have that:*

a. $p \in [0, p^-]$, *if* $\beta^- \le 1 \le \beta^+$, *with* $\beta^- \beta^+ > 1$, *or* $1 \le \beta^- \le \beta^+$, *or* $\gamma_2^{max} = \gamma_3^{max}$;

b. $p \in [0, p^-] \cup [p^+, 1]$, *if* $\beta^- \le 1 \le \beta^+$, *with* $\beta^- \beta^+ = 1$;

c. $p \in [p^+, 1]$, *if* $\beta^- \le 1 \le \beta^+$, *with* $\beta^- \beta^+ < 1$ *or* $\beta^- \le \beta^+ \le 1$;

2. *Case* $s_1 = -1 = -s_2$, *we have that:*
*for* $J_1(T)$, $p \in [p^+, 1]$;
*for* $J_6(T)$:

a. $p \in [p^+, 1]$, *if* $\beta^- \le 1 \le \beta^+$, *or* $\beta^- \le \beta^+ \le 1$;

b. $p \in [0, p^-]$ *or* $p \in [p^+, 1]$, *if* $1 \le \beta^- \le \beta^+$, *or* $\gamma_2^{max} = \gamma_3^{max}$;

3. *Case* $s_1 = +1 = -s_2$, *we have that:*
*for* $J_1(T)$, $p \in [0, p^-]$;
*for* $J_6(T)$:

a. $p \in [0, p^-]$, *if* $\beta^- \le 1 \le \beta^+$, *or* $1 \le \beta^- \le \beta^+$, *or* $\gamma_2^{max} = \gamma_3^{max}$;

b. $p \in [0, p^-]$ *or* $p \in [p^+, 1]$, *if* $\beta^- \le \beta^+ \le 1$;

4. *Case* $s_1 = s_2 = -1$, *we have that:*
*for* $J_1(T)$:

a. $p \in \left[0, p^-\right]$, *if* $\beta^- \le 1 \le \beta^+$, *with* $\beta^- \beta^+ < 1$, *or* $\beta^- \le \beta^+ \le 1$;

b. $p \in \left[0, p^-\right] \cup \left[p^+, 1\right]$, *if* $\beta^- \le 1 \le \beta^+$, *with* $\beta^- \beta^+ = 1$;

c. $p \in \left[p^+, 1\right]$, *if* $\beta^- \le 1 \le \beta^+$, *with* $\beta^- \beta^+ > 1$, *or* $1 \le \beta^- \le \beta^+$, *or* $\gamma_2^{\max} = \gamma_3^{\max}$;

*for* $J_6(T)$:

a. $p \in \left[0, p^-\right]$ *or* $p \in \left[p^+, 1\right]$, *if* $\beta^- \le 1 \le \beta^+$, *with* $\beta^- \beta^+ > 1$, *or* $\beta^- \beta^+ < 1$, *or* $\gamma_2^{\max} = \gamma_3^{\max}$;

b. $p \in \left[0, p^-\right] \cup \left[p^+, 1\right]$, *if* $\beta^- \le 1 \le \beta^+$, *with* $\beta^- \beta^+ = 1$;

c. $p \in \left[p^+, 1\right]$, *if* $\beta^- \le \beta^+ \le 1$;

d. $p \in \left[0, p^-\right]$, *if* $1 \le \beta^- \le \beta^+$;

*where* $\beta^- = \dfrac{\gamma_3^{\max} - \gamma_1^{\max}}{\gamma_1^{\max}}, \beta^+ = \dfrac{\gamma_2^{\max}}{\gamma_3^{\max} - \gamma_2^{\max}}, p^+ = \dfrac{\gamma_1^{\max}}{\gamma_3^{\max}}, p^- = \dfrac{\gamma_3^{\max} - \gamma_2^{\max}}{\gamma_3^{\max}}$.

The functionals $J_1$ and $J_6$ are maximized for $p = 0$ or $p = 1$ if $\gamma_1^{\max} = \gamma_2^{\max} = \gamma_3^{\max}$.

The optimization algorithms are tested on Re di Roma square, a part of the urban network of Rome and on Via Parmenide crossing, a little network in Salerno (Italy).

We consider approximations obtained by the numerical method of Godunov, with space step $\Delta x = 0.01$ and time step determined by the CFL condition. The road network is simulated in a time interval $[0,T]$, where $T = 30$ min. As for the initial conditions on the roads of the network, we assume that, at the starting instant of simulation ($t = 0$), all roads are empty. We studied different simulation cases: right of way parameters, that optimize the cost functionals (*optimal case*); random right of way parameters (*static random case*), i.e. the right of way parameters are chosen in a random way at the beginning of the simulation process; fixed right of way parameters (*fixed case*), the same for each junction; dynamic random parameters (*dynamic random case*), i.e. right of way parameters change randomly at every step of the simulation process.

Re di Roma square is a big traffic circle with 12 roads (6 entering roads and 6 exiting ones), 6 junctions of 2 × 1 type and 6 junctions of 1 × 2 type. In Figure 2 (left), the topology of Re di Roma Square is reported, with junctions of 2×1 type (1, 3, 5, 7, 9, 11) in white, and junctions of 1 × 2 type (2, 4, 6, 8, 10, 12) in black. The traffic distribution coefficients at 1 × 2 junctions are determined by the road capacities (and the characteristics of the nearby portion of the Rome urban network). Therefore we focused on the right of way parameters for the 2 × 1 junctions, whose choice corresponds to the use of yielding and stop signs, or to the regulation of red and green phases for traffic lights.

We assume boundary conditions 0.3 for roads with non infinite endpoints and we choose for the fixed case $p = 0.2$, the mean value of the static random simulations. The simulative results present some expected features and some unexpected ones. The performances of the optimal and dynamic random coefficients are definitely superior with respect to the other two. However, performances are surprisingly good, taking into account that the optimal choice was obtained by local optimization and asymptotic state, and that the dynamic random result is very close to the optimal one. Such behavior is clear for $J_1$ functional (see Figure 3), and even more marked for $J_2$ functional, which explodes for the static random and fixed parameters in case of high traffic load (see Figure 4). The explanation for such

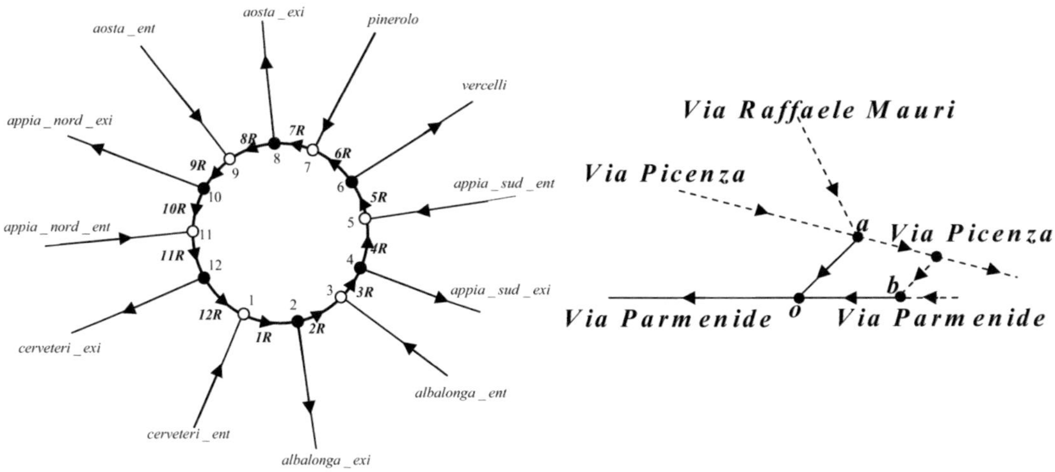


Fig. 2. Topology of Re di Roma Square (left) and Via Parmenide crossing in Salerno (right).

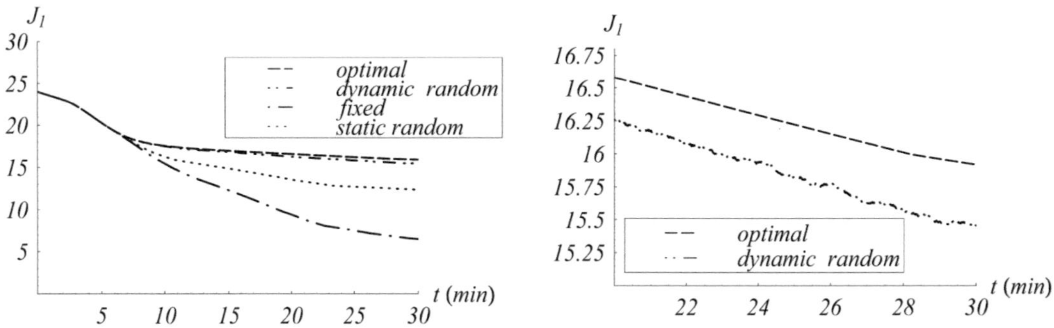


Fig. 3. $J_1$ simulated with zero initial conditions for all roads, boundary conditions equal to 0.3 (left) and zoom around the optimal and dynamic random case (right).

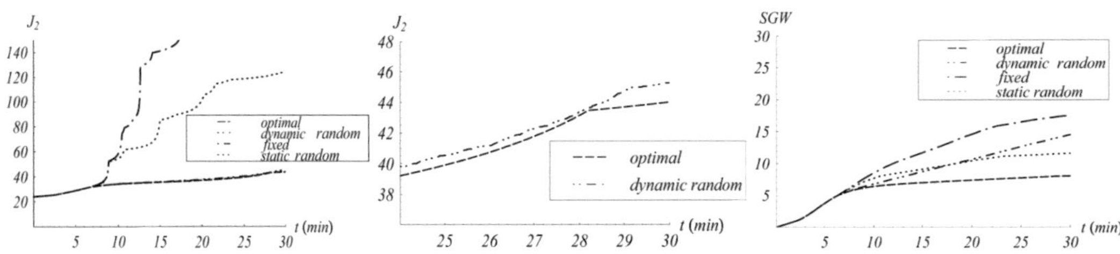


Fig. 4. $J_2$ simulated with zero initial conditions for all roads, boundary conditions equal to 0.3 (left) and zoom around the optimal and dynamic random case (central). Behavior of the *SGW* functional in the case of boundary conditions equal to 0.3 and same initial conditions for all roads (right).

explosion is the following: in some situation, the traffic circle gets completely stuck, thus the travelling time tends to infinity. This fact is also confirmed by the behavior of the cars densities on the roads, which are very irregular in the dynamic random simulations. When we consider networks with a great number of nodes, the time average of optimal parameters

can approach 0.5, and this justifies the similarities among dynamic random simulations and optimal ones. Hence, to discriminate between them, it is necessary to consider the *SGW* functional. The latter is very high for the dynamic random case and very low for the optimal one (even less than the fixed or static random case), see Figure 4 (right).
Then, we analyzed a small area of the Salerno urban network, characterized by congestion due to a traffic light, that presents a cycle with a red phase too long. In particular, in Figure 2 (right) the portion of the interested area is depicted. We focus on the crossing indicated by $o$. The incoming road from point $a$ to point $o$ (a portion of Via Mauri, that we call road $a - o$) is very short and connects Via Picenza to Via Parmenide. The incoming road from point $b$ to point $o$ (that we call road $b$–$o$) is a part of Via Parmenide. Crossing $o$ is ruled by a traffic light, with a cycle of 120 seconds, where the phase of green is 15 seconds for drivers, who travel on the road $a - o$. It is evident that such situation leads to very high traffic densities on the road $a - 0$ as the little duration of green phase does not always allow to absorb queues. From a probabilistic point of view, we can say that road $b - o$ has a right of way parameter, that is: $p = \frac{105}{120} = 0.875$, while road $a - o$ has a right of way parameter $q = 1 - p = 0.125$.
This particular crossing has been studied in order to understand how to improve the conditions of traffic in presence of a traffic light. We considered a boundary data 0.8 for roads, that enter the junction $o$ and a boundary condition 0.3 for the outgoing road. Figure 5 (left) reports the functional $J_1$ in optimal, dynamic random and fixed cases. It is evident that the optimal case is higher than the fixed simulation (that corresponds to the real case $p$ = 0.875); hence, actually, Via Parmenide in Salerno does not follow a traffic optimization policy. In fact there are some time intervals in which cars are stopped by the traffic light, while other roads are completely empty. This situation means that the cycle of the traffic light is too long. A solution could be to reduce the cycle or substitute the traffic light with a stop sign.
Simulations show that dynamic random algorithms and optimization approaches are totally different for Via Parmenide, respect to Re di Roma square. This is due to the nature itself of the dynamic random simulation (Figure 5, right), that is similar to a fixed case with $p = 0.5$, which is the minimum for $J_1$. For Via Parmenide, only one traffic parameter is used, whose analytical optimization gives a solution far from 0.5; hence, the dynamic random simulation and optimal ones cannot be similar.

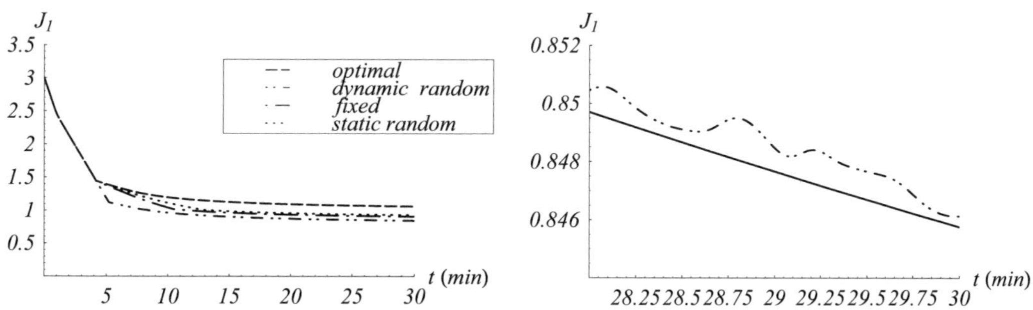


Fig. 5. Left: behavior of $J_1$ for Via Parmenide crossing. Right: behavior of $J_1$ for Via Parmenide crossing among the dynamic random simulation (dot dot dashed line) and the fixed simulation with $p = 0.5$ (solid line).

The optimizations of local type, like the ones that we are considering here, could not necessarily imply global performance improvements for big networks. Table 1 reports the distribution coefficients used for simulations of Re di Roma Square. Roads cerveteri_exi, albalonga_exi, appia_sud_exi, vercelli, aosta_exi, appia_nord_exi have a boundary data equal to 0.4, while the other ones equal to 0.35.

| Junction $i$ | $\alpha_{iR,(i-1)R}$ | $\alpha_{i,(i-1)R}$ |
|---|---|---|
| $i = 2$ | 0.866071 | 0.133929 |
| $i = 4$ | 0.459854 | 0.540146 |
| $i = 6$ | 0.800971 | 0.199029 |
| $i = 8$ | 0.730612 | 0.269388 |
| $i = 10$ | 0.536050 | 0.463950 |
| $i = 12$ | 0.753927 | 0.246073 |

Table 1. Traffic distribution parameters for junctions of 1 ×2 type in Re di Roma Square.

We show that for the chosen initial data the algorithm for the maximization of velocity assures globally the best performance for the network, also in terms of average times, and kinetic energy (see Figures 6 and 7). The goodness of opt$J_1$ for global performances is confirmed by the behavior of $J_2$. In fact, opt$J_2$ $J_7$ and opt$J_6$ can let $J_2$ explode, i.e. the traffic circle is stuck and the time to run inside goes to infinity. This situation is more evident in the total kinetic energy, $J_6$, which tends to zero when opt$J_1$ is not used. This means that the cars flux is going to zero, as evident from Figure 8 (left), hence roads inside the circle are becoming full. A consequence of this phenomenon is also visible in $J_7$ evolution, that tends to infinity.

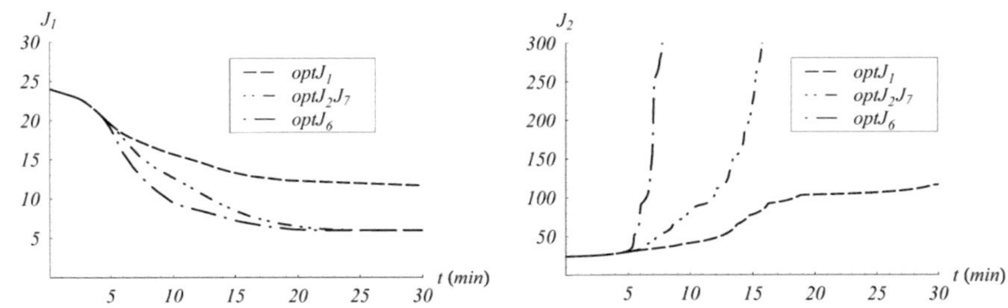


Fig. 6. Behavior of $J_1$ (left) and $J_2$ (right), using the parameter $p$ which optimizes $J_1$, opt$J_1$, $J_2$ and $J_7$, opt$J_2$ $J_7$, and $J_6$, opt$J_6$.

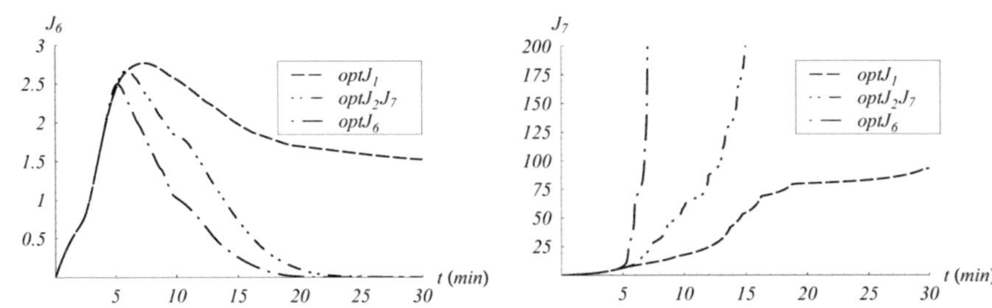


Fig. 7. Behavior of $J_6$ (left) and $J_7$ (right), using the parameter $p$ which optimizes $J_1$, opt$J_1$, $J_2$ and $J_7$, opt$J_2$ $J_7$, and $J_6$, opt$J_6$.

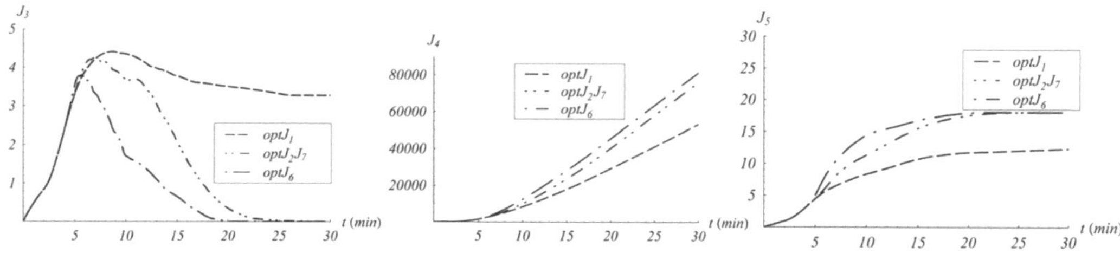


Fig. 8. Behavior of the optimal cost functionals $J_3$ (left), $J_4$ (central) and *SGW* (right).

Observe that the amount of traffic load, visible in $J_4$ (Figure 8 (central)), tends to decrease using the priority parameter which maximizes $J_1$. Moreover, the behavior of $J_5$ (Figure 8 (right)), that measures the velocity variation, indicates that the use of opt$J_1$ leads to more regular densities on roads, giving advantages in terms of security. Remember that there are no optimization algorithms for the functionals $J_k$, $k$ =3, 4, 5 and they are computed directly using opt$J_1$, opt$J_2$ $J_7$ and opt$J_6$.

### 4.2 Optimization of junctions of 1 ×2 type

We focus on junctions of 1 ×2 type, labelling with 1 the incoming road and with 2 and 3 the outgoing ones and we consider $\alpha$ as control. For more details, see Cascone et al. (2008 a).

**Theorem 8** *Consider a junction J of* 1 ×2 *type and T sufficiently big. The cost functionals* $J_1(T)$, $J_2(T)$, $J_6(T)$ *and* $J_7(T)$ *are optimized for* $\alpha = \frac{1}{2}$, *with the exception of the following cases (in some of them, the optimal control does not exist but is approximated ):*

1. *if* $\gamma_2^{\max} < \gamma_1^{\max} \le \gamma_3^{\max}$ *and* $\gamma_2^{\max} < \dfrac{\gamma_1^{\max}}{2}$, $\alpha = \alpha_2$;
2. *if* $\gamma_2^{\max} \le \gamma_3^{\max} \le \gamma_1^{\max}$ *and* $\gamma_3^{\max} \le \dfrac{\gamma_1^{\max}}{2}$, $\alpha = \alpha_1 + \varepsilon$;
3. *if* $\gamma_2^{\max} \le \gamma_3^{\max} \le \gamma_1^{\max}$, $\alpha = \dfrac{\gamma_2^{\max}}{\gamma_2^{\max} + \gamma_3^{\max}}$;
4. if $\gamma_3^{\max} \le \gamma_2^{\max} < \gamma_1^{\max}$, *we have to distinguish three cases:*

- *if* $\alpha_2 = \frac{1}{2}$, $\alpha = \frac{1}{2} - \varepsilon$;
- *if* $\frac{1}{2} \le \alpha_1 < \alpha_2$, $\alpha = \alpha_1 + \varepsilon$;
- *if* $\alpha_1 < \alpha_2 \le \frac{1}{2}$, $\alpha = \alpha_2 - \varepsilon$;

*where* $\alpha_1 = 1 - \dfrac{\gamma_3^{\max}}{\gamma_1^{\max}}$, $\alpha_2 = \dfrac{\gamma_2^{\max}}{\gamma_1^{\max}}$ *and* $\varepsilon$ *is small and positive.*

We present simulation results for a road network, that consists of 6 junctions of 1×2 and 2×2 type, see Figure 9.

For every junction of 2 ×2 type, we set all the entries of the distribution matrix *A* equal to 0.5. Hence, no control is considered for such junctions. The network is simulated in such conditions: initial data equal to 0.3 for all roads at the starting instant of simulation ($t$ = 0); boundary data of Dirichlet type, equal to 0.45 for road $a_1$, while for roads $a_3$, $c_2$, $f_1$, and $f_2$, we

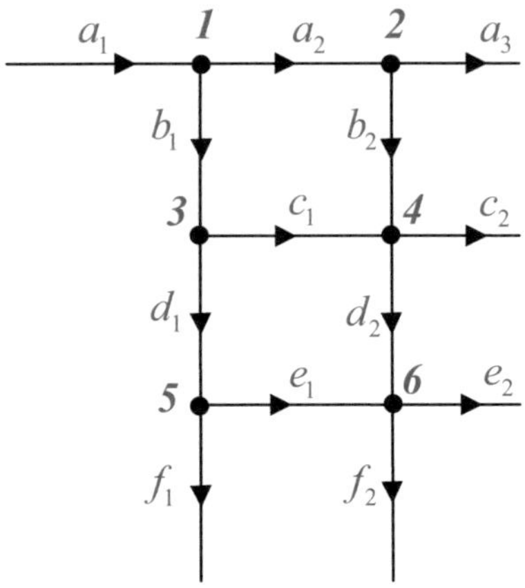


Fig. 9. Topology of the network.

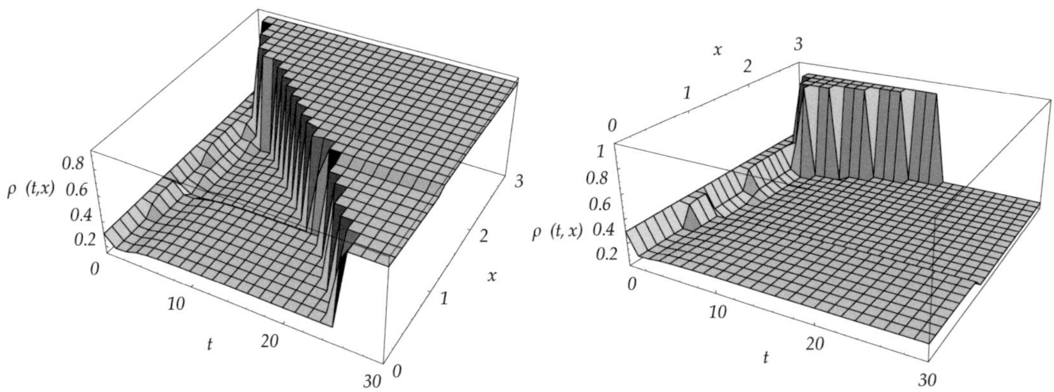


Fig. 10. Left. Density $\rho(t, x)$ on roads $b_1$ (from 0 to 1 on the axis $x$), $d_1$ (from 1 to 2 on the axis $x$) and $f_1$ (from 2 to 3 on the axis $x$), with $\alpha = 0.2$ for $1 \times 2$ junctions. Right. Density $\rho(t, x)$ on roads $b_1$ (from 0 to 1 on the axis $x$), $d_1$ (from 1 to 2 on the axis $x$) and $f_1$ (from 2 to 3 on the axis $x$), with optimal distribution coefficients for $1 \times 2$ junctions.

choose a Dirichlet boundary data equal to 0.9; time interval of simulation $[0,T]$, where $T = 30$ min. We analyzed traffic conditions for different values of $\alpha$. Figure 10 (left) reports the density $\rho(t, x)$ on roads $b_1$, $d_1$ and $f_1$ for $\alpha = 0.2$, assuming that all roads have length equal to 1. High levels of density interest these vertical roads, hence they tend to be more heavily congested than others. This can be seen in Figure 10 where, at $t = 10$, the road $f_1$ is already congested with a density value almost equal to 0.9. At $t = 25$, the intense traffic of roads $f_1$ propagates backward and influences roads $b_1$ and $d_1$. The traffic densities on other roads is very low.

When we deal with the optimal choice of the distribution coefficients, densities on roads $c_2$, $e_2$, $f_1$ and $f_2$ tend to increase. However, the optimal choice better redistributes traffic flows on the whole network, as we can see from Figure 10 (right), that shows the car density $\rho(t, x)$ for roads $b_1$, $d_1$ and $f_1$.

Then, we compared three scenarios ($\alpha = 0.2$, $\alpha = 0.8$ and optimal $\alpha$). We concluded that a real decongestion effect is evident for optimal distribution coefficients (see Figure 11, that shows the cost functionals $J_1$ and $J_2$). This phenomenon is also evident for the behaviors of $J_6$ and $J_7$, see Figure 12. In fact, the kinetic energy, represented by $J_6$, tends to zero when congestion problems are evident, as in the case $\alpha = 0.2$. This means that the cars flux is going to zero and roads of the network are becoming to be full. An improvement of car traffic is obtained for $\alpha = 0.8$ but the better situation in terms of viability is always reached in the optimal case. Indeed, for the cost functional $J_7$, the presence of decongestion phenomena is more evident when networks parameters are not the optimal ones. In fact, fixing $\alpha = 0.2$ and $\alpha = 0.8$, $J_7$ tends to infinity for big times, meaning that the velocity of cars is decreasing, with consequent filling of roads.

The dynamic random simulation follows the behavior of the optimal one, as we can see from Figure 13 (left). One could ask if an optimization is necessary, since random choices leads to similar functional values. The dynamic random simulation, in the reality of urban networks, implies that drivers flow is very chaotic, since drivers choices rapidly change during their own travel. Let us show this phenomenon considering the Stop and Go Waves functional (*SGW*).

Figure 13 (right) shows a great variation of velocity for the dynamic random choice, which implies a higher probability of car accidents. Note that the optimal case for *SGW* is simulated according to the optimization algorithm for the cost functionals $J_1$ and $J_2$ (and not for *SGW* itself).

From a statistical point of view, it is possible to understand why dynamic random simulations are very similar to the optimal case for functionals $J_1$ and $J_2$. From Theorem 8, the optimal choice for the distribution coefficient is almost always 0.5, and this is the expected average value of random choices.

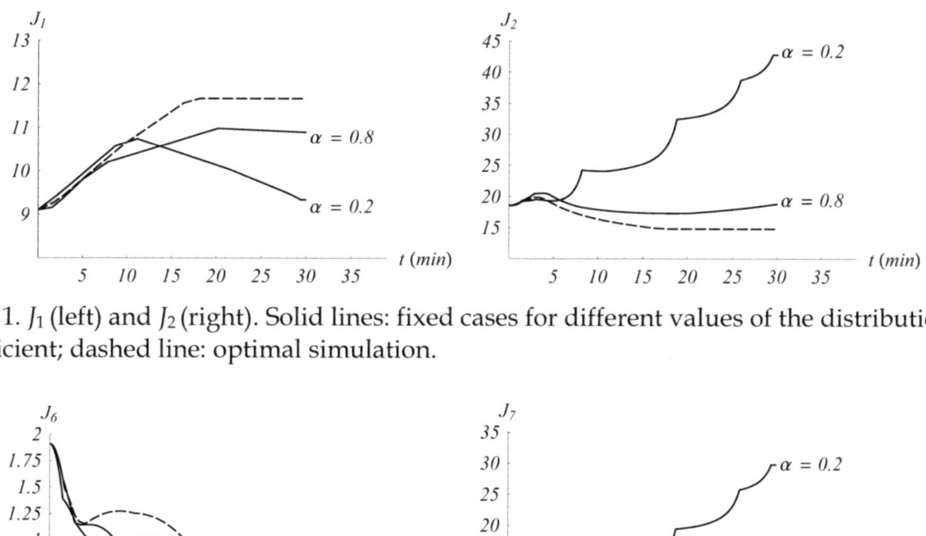


Fig. 11. $J_1$ (left) and $J_2$ (right). Solid lines: fixed cases for different values of the distribution coefficient; dashed line: optimal simulation.

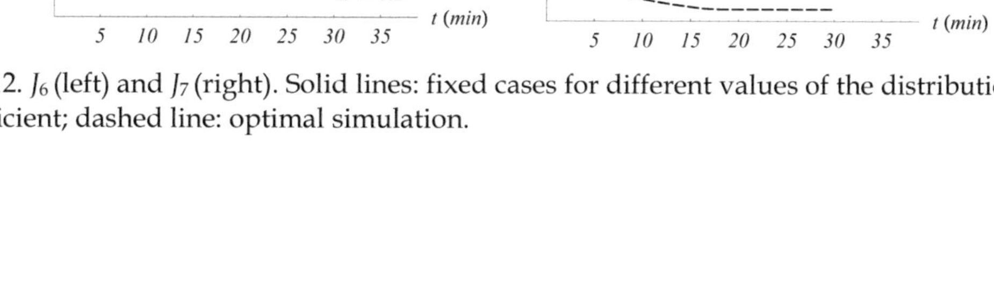


Fig. 12. $J_6$ (left) and $J_7$ (right). Solid lines: fixed cases for different values of the distribution coefficient; dashed line: optimal simulation.

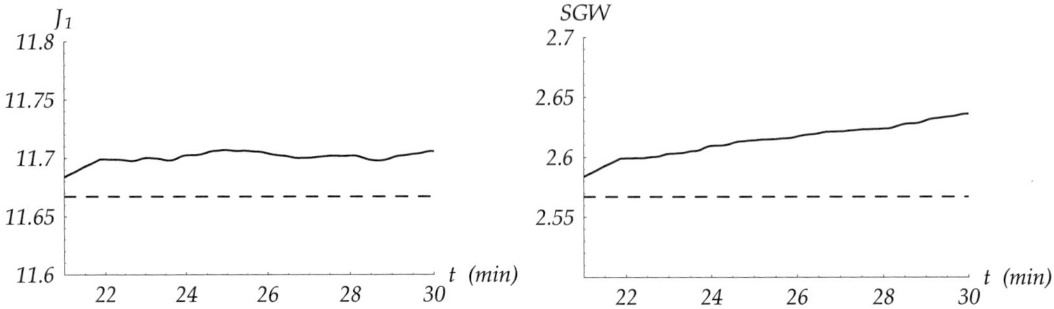


Fig. 13. Left: comparison among the dynamic random simulation and the optimal case for $J_1$. Right: behavior of *SGW* in the optimal and in the dynamic random case; dashed line: optimal simulation.

## 5. Optimal distribution of traffic flows at junctions in emergency cases

The problem we face here is to find the values of traffic distribution parameters at a junction in order to manage critical situations, such as car accidents. In this case, beside the ordinary cars flows, other traffic sources, due to emergency vehicles, are present. More precisely, assume that a car accident occurs on a road of an urban network and that some emergency vehicles have to reach the position of the accident, or of a hospital.
We define a velocity function for such vehicles:

$$\omega(\rho) = 1 - \delta + \delta v(\rho), \tag{6}$$

with $0 < \delta < 1$ and $v(\rho)$ as in (4). Since $\omega(\rho_{\max}) = 1 - \delta > 0$, it follows that the emergency vehicles travel with a higher velocity with respect to cars. Notice that (6) refers to the previous formula coincides with the velocity of the ordinary traffic for $\delta = 1$.
Consider a junction *J* with 2 incoming roads and 2 outgoing ones. Fix an incoming road $I_\varphi$, $\varphi = 1, 2$, and an outgoing road $I_\psi$, $\psi = 3, 4$. Given an initial data $(\rho_{\varphi,0}, \rho_{\psi,0})$, we define the cost functional $W_{\varphi,\psi}(t)$, which indicates the average velocity of emergency vehicles crossing $I_\varphi$ and $I_\psi$:

$$W_{\varphi,\psi}(t) = \int_{I_\varphi} \omega(\rho_\varphi(t,x))dx + \int_{I_\psi} \omega(\rho_\psi(t,x))dx.$$

For a fixed time horizon $[0,T]$, the aim is to maximize $\int_0^T W_{\varphi,\psi}(t)$ by a suitable choice of the traffic distribution parameters $\alpha_{\psi,\varphi}$ for $T$ sufficiently big.
Assigned the path consisting of roads 1 and 3, the cost functional $W_{1,3}(T)$ is optimized choosing the distribution coefficients according to the following theorem (for more details see Manzo et al. (2010)).
**Theorem 9** *For a junction J of* $2 \times 2$ *type and T sufficiently big, the cost functional* $W_{1,3}(T)$ *is maximized for* $\alpha = 1 - \frac{\gamma_4^{\max}}{\gamma_1^{\max}}$, $0 \le \beta < 1 - \frac{\gamma_4^{\max}}{\gamma_1^{\max}}$, *with the exception of the following cases, where the optimal controls do not exist but are approximated by:*

- $\alpha = \varepsilon_1,\ \beta = \varepsilon_2,\ if\ \gamma_1^{\max} \le \gamma_4^{\max};$

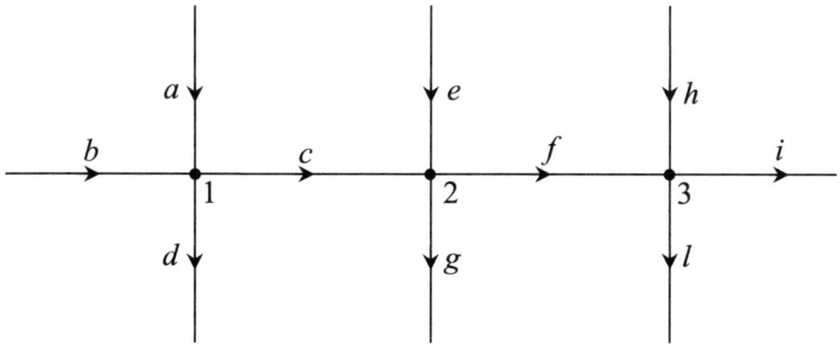


Fig. 14. Topology of the cascade junction network.

- $\alpha = \frac{\gamma_3^{\max}}{\gamma_3^{\max} + \gamma_4^{\max}} - \varepsilon_1,\ \beta = \frac{\gamma_3^{\max}}{\gamma_3^{\max} + \gamma_4^{\max}} - \varepsilon_2,\ if\ \gamma_1^{\max} > \gamma_3^{\max} + \gamma_4^{\max},$

  *for* $\varepsilon_1$ *and* $\varepsilon_2$ *small, positive and such that* $\varepsilon_1 \neq \varepsilon_2$.

Consider the network in Figure 14, described by 10 roads, divided into two subsets, $R_1 = \{a, d, e, g, h, l\}$ and $R_2 = \{b, c, f, i\}$ that are, respectively, the set of inner and external roads. Assuming that the emergency vehicles have an assigned path, we analyze the behavior of the functional:

$$W(t) = W_{ac}(t) + W_{ef}(t) + W_{hi}(t).$$

The evolution of traffic flows is simulated using the Godunov scheme with $\Delta x = 0.0125$, and $\Delta t = \frac{\Delta x}{2}$ in a time interval $[0,T]$, where $T = 100$ min. Initial and boundary data are chosen in order to simulate a network with critical conditions on some roads, as congestions due to the presence of accidents (see Table 2).

| Road | Initial condition | Boundary data |
|---|---|---|
| *a* | 0.1 | 0.1 |
| *b* | 0.65 | 0.65 |
| *c* | 0.75 | / |
| *d* | 0.95 | 0.95 |
| *e* | 0.2 | 0.2 |
| *f* | 0.65 | / |
| *g* | 0.95 | 0.95 |
| *h* | 0.25 | 0.25 |
| *i* | 0.55 | 0.55 |
| *l* | 0.95 | 0.95 |

Table 2. Initial conditions and boundary data for roads of the cascade junction network.

Figure 15 shows the temporal behavior of $W(t)$ measured on the whole network. As we can see, the optimal cost functional is higher than the random ones, hence the principal aim is achieved for the chosen data set. Notice that, in general, optimal global performances on networks could also not be achieved, as the traffic state is strictly dependent on initial and boundary data.

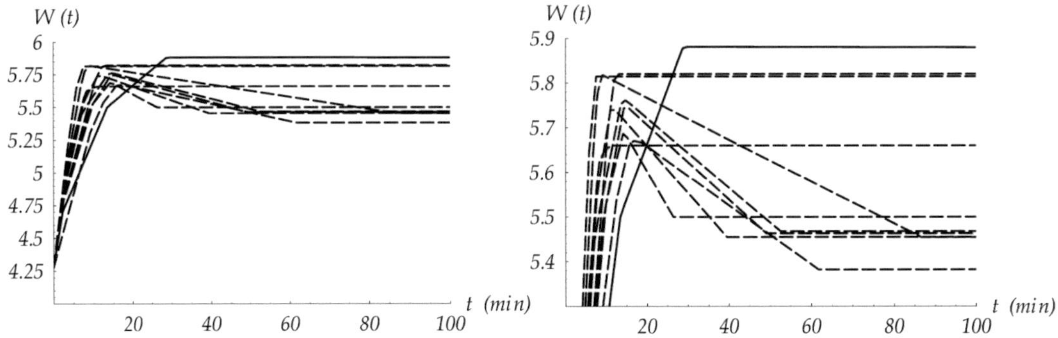


Fig. 15. Evolution of $W(t)$ for optimal choices (continuous line) and random parameters (dashed line); left: behavior in $[0,T]$; right: zoom around the asymptotic values.

## 6. Conclusions

Traffic regulation techniques for the optimization of car traffic flows in congested urban networks was considered. The approach used for the description of traffic flows is of fluid-dynamic type. The main advantages of this approach, with respect to existing ones, can be summarized as follows. The fluid-dynamic models are completely evolutive, thus they are able to describe the traffic situation of a network at every instant of time. This overcomes the difficulties encountered by many static models. An accurate description of queues formation and evolution on the network is possible. The theory permits the development of efficient numerical schemes also for very large networks. This is possible since traffic at junctions is modelled in a simple and computationally convenient way (resorting to a linear programming problem). The performance analysis of the networks was made through the use of different cost functionals, measuring car average velocity weighted or not weighted with the number of cars moving on each road, the average travelling time, velocity variation, kinetic energy, etc. Exact analytical results were given for simple junctions of $1 \times 2$ and $2 \times 1$ type, and then used in order to simulate more complex urban networks. Moreover the problem of emergency vehicles transit has been treated. The problem has been faced choosing a route for emergency vehicles (not dedicated, i.e. not limited only to emergency needs) and redistributing traffic flows at junctions on the basis of the current traffic load in such way that emergency vehicles could travel at the maximum allowed speed along the assigned roads (and without blocking the traffic on other roads). All the optimization results have been obtained using a decentralized approach, i.e. an approach which sets local optimal parameters for each junction of the network. In future we aim to extend the optimization results to more general junctions and to explore global optimization techniques. In addition, the definition and optimization of functionals which take into account the emission and propagation of pollutants produced by cars might provide powerful technological tools to rationalize the design and use of public and private transportation resources, and to reduce unpleasant effects of urban traffic on the environment.

## 7. References


Aw, A. & Rascle, M. (2000). Resurrection of "second order" models of traffic flow?, *SIAM J. Appl. Math.*, Vol. 60, pp. 916-938.

Bardos, C.; Le Roux, A.Y. & Nedelec, J.C. (1979). First Order Quasilinear Equations with Boundary Conditions, *Commun. Partial Differential Equations*, Vol. 4, pp. 1017-1034.

Bressan, A. (2000). *Hyperbolic Systems of Conservation Laws - The One-dimensional Cauchy Problem*, Oxford Univ. Press.

Cascone, A.; D'Apice, C.; Piccoli, B. & Rarità, L. (2007). Optimization of traffic on road networks, *Mathematical Models and Methods in Applied Sciences*, Vol. 17, No. 10, pp. 1587-1617.

Cascone, A.; D'Apice, C; Piccoli, B. & Rarità, L. (2008). Circulation of car traffic in congested urban areas, *Communication in Mathematical Sciences*, Vol. 6, No. 3, pp. 765-784.

Cascone A.; Manzo, R.; Piccoli B. & Rarità L. (2008). Optimal vs randomness for car traffic regulation, *Physical Review E, Statistical, Nonlinear, And Soft Matter Physics*, Vol. 78, No. 2, 026113(5).

Chitour Y. & Piccoli B. (2005). Traffic circles and timing of traffic lights for cars flow, *Discrete and Continous Dynamical Systems-Series B*, Vol. 5, pp. 599-630.

Coclite, G.M.; Garavello, M. & Piccoli, B. (2005). Traffic flow on a road network, *SIAM J. Math. Anal.*, Vol. 36, No. 6, pp. 1862-1886.

Colombo, R.M. (2002). Hyperbolic phase transitions in traffic flow, *SIAM J. Appl. Math.*, Vol. 63, No. 2, pp. 708-721.

Colombo, R.M. (2002). On a 2x2 hyperbolic traffic flow model, *Math. Comput. Modelling*, Vol. 35, pp. 683-688.

Cutolo, A.; D'Apice, C. & Manzo, R. (2009). Traffic optimization at junctions to improve vehicular flows, *Preprint D.I.I.M.A.*, No. 57.

Dafermos, C. (1999). Hyperbolic Conservation Laws in Continuum Physics, Springer-Verlag.

Daganzo, C. (1995). Requiem for second-order fluid approximation to traffic flow, *Transportation Res., Part B*, Vol. 29, pp. 277-286.

Garavello, M. & Piccoli, B. (2006). Traffic Flow on Networks, Applied Math Series Vol. 1, American Institute of Mathematical Sciences.

Garavello, M. & Piccoli, B. (2006). Traffic flow on a road network using the Aw-Rascle model, *Commun. Partial Differential Equations*, Vol. 31, No. 2, pp. 243-275.

Greenberg, J.M.; Klar, A. & Rascle, M. (2003). Congestion on multilane highways, *SIAM J. Appl. Math.*, Vol. 63, No. 3, pp. 818-833.

Gugat, M.; Herty, M.; Klar A. & Leugering, G. (2005). Optimal Control for Traffic Flow Networks, *Journal of Optimization Theory and Applications*, Vol. 126, pp. 589-616.

Helbing, D. (2001). Traffic and related self-driven many-particle systems, *Reviews of Modern Physics*, Vol. 73, pp. 1067-1141.

Helbing, D.; Lammer, S. & Lebacque, J. P. (2005). Self-organized control of irregular or perturbed network traffic, *Optimal Control and Dynamic Games, C. Deissenberg and R.F. Hartl eds., Springer, Dordrecht*, pp. 239-274.

Herty, M. & Klar, A. (2003). Modelling, Simulation and Optimization of Traffic Flow Networks, *SIAM J. Sci. Comp.*, Vol. 25, pp. 1066-1087.

Holden, H. & Risebro N.H. (1995). A mathematical model of traffic flow on a network of unidirectional roads, *SIAM J. Math. Anal.*, Vol. 26, pp. 999-1017.

Lighthill, M.J. & Whitham, G.B. (1955). On kinetic waves. II. Theory of traffic flows on long crowded roads, *Proc. Roy. Soc. London Ser. A*, Vol. 229, pp. 317-345.
Manzo, R.; Piccoli, B. & Rarità, L. (2010). Optimal distribution of traffic flows at junctions in emergency cases, *European Journal on Applied Mathematics*, submitted.
Payne, H.J. (1971). Models of freeway traffic and control, *Simulation Council Proc.*, Vol. 1, pp. 51-61.
Payne, H.J. (1979). FREFLO: a macroscopic simulation model of freeway traffic, *Transportation Research Record*, Vol. 722, pp. 68-75.
Richards, P.I. (1956). Shock waves on the highway, *Oper. Res.*, Vol. 4, pp. 42-51.
Serre, D. (1996). Systems of conservation laws I and II, Diderot Editeur, Paris.
Whitham, G.B. (1974). Linear and Nonlinear Waves, Pure and Applied Math., Wiley-Interscience, New York.

# 7

# Dynamic Modelling and Simulation of Electrochemical Energy Systems for Electric Vehicles

Lucia Gauchia and Javier Sanz
*Carlos III University, Electric Engineering Department*
*Spain*

## 1. Introduction

The current energy scenery is dominated by fossil fuels, especially oil. This dependency is turning critical due to the reducing reserves, uncertain oil resources, and political and economical ramifications of a concentration of fossil fuel reserves in a limited number of regions. The transportation sector is especially affected by this situation and needs to develop new energy vectors and systems to reduce the oil dependency whilst attending to environmental issues. Therefore, vehicle manufacturers are turning to hybrid and electric vehicles. Hybrid vehicles combine an internal combustion engine (ICE) with energy storage systems, which allows reducing the installed power of the ICE, and consequently the fuel consumption and pollutant emissions. With this power train, the user is capable of driving in a pure electric mode, through the energy storage system (normally batteries), or in a hybrid mode with both ICE and storage for more challenging driving cycles.

Electric vehicles are especially interesting due to the exclusion of the ICE, which reduces to zero the emissions, and presents a higher efficiency of the power train and environmentally friendly operation. However, even if these reasons are activating its interest, there are several drawbacks which should be solved before reaching a mass production scale. Some of these issues include the development of energy technologies able to guarantee an adequate vehicle range, attractive power ratings and safe, simple and fast recharge.

Nowadays there is no electric energy storage technology which can exhibit both high energy and power densities, necessary to meet range and accelerating requirements. Therefore, there is an intensive research to develop new materials for electrochemical energy devices and to hybridize electrochemical energy systems to reach the necessary power and energy specifications. The most popular technologies are Ni-Mh and Li-based batteries, which present higher energy densities than classic Pb-acid batteries. However, these technologies cannot achieve the range obtained with fossil fuels. Therefore, other energy systems, such as fuel cells or flow batteries are being studied as part of a hybrid electric vehicle power train.

Finally, this energy system research should be done taking into account the particular situation of transportation, where the weight, volume and cost of the systems included are relevant for a successful and massive use of the electric vehicle. To carry out this research in the final application stage of electrochemical systems, it is necessary to be able to test, model and simulate this system in real operating conditions.

Electric vehicle development can be highly expensive and complex if the power train is considered as a whole during laboratory tests, when the objectives are the evaluation of alternative designs. Different testing approaches can be used for this purpose: from the pure simulation to the real hardware testing. The pure simulation can present some problems, such as non-realistic scenarios or excessive simplification which can later provoke problems in the integration of the real system. On the other hand, real hardware testing can be expensive, even if highly accurate. That is why intermediate approaches, such as hardware-in-the-loop (HIL) simulation, in which some hardware elements are integrated into a real-time simulation, can be very recommendable.

This chapter presents a HIL simulation platform for hybrid electric vehicles, taking into account different technologies, studying how the control strategy selected affects each of the systems involved.

## 2. Testing methods

Most testing methods can be classified as:

1. **Pure simulation:** The testing of the complete system is done exclusively through software. Therefore, the cost is relatively low as the resources needed are computers and specific software to carry out the simulation. Even if the programming can present its own problems, in general, this procedure is simpler than the rest of approaches, but is also less accurate due to the impossibility to compare it to real measurements.
2. **Hardware-in-the-loop (HIL):** In the HIL simulation (Maclay, 1997) part of the system is simulated and part is real hardware. This increases the accuracy and adds real measures to the pure simulation previously mentioned. Moreover, no highly detailed models have to be included nor any highly complex hardware system is used. Of course, it also increases the complexity due to the hardware inclusion; however, as hardware is only one part of the system it is still feasible without increasing excessively cost and complexity. It is especially useful for hybrid energy sources and electric drives, as explained and classified by Bouscayrol (Bouscayrol, 2008). This classification is done considering the type of signal which interacts between the simulated and hardware system, as shown in Fig. 1:
   a. **Signal level HIL:** The first type is the signal level HIL simulation. As seen in Fig. 1 the hardware element is a control scheme or similar, which interacts through control signals with the simulated environment (electric machine, mechanical load or power electronics). In this case, a control scheme can be tested without actually setting up a complex and expensive laboratory test bench due to the fact that the control schemes only need a processor and computer. This approach has been applied, e.g. for power electronics testing (Lu et al., 2007).
   b. **Power level HIL:** In this case one of the simulated systems is substituted by the hardware real system, whilst the rest remains simulated. Now the simulated and physical system also require power signals, as shown in Fig. 1. This approach is being increasingly used for highly complex systems, such as vehicles, electric drives or even ships (Ren et al., 2008).
   c. **Mechanical level:** At this level the whole drive (control, power electronics and electric machine) is hardware. The simulated load is setup with a mechanical load or another electric machine. Therefore, both hardware and simulated elements

interact mechanically, through the shaft, as depicted in Fig. 1. This mechanical level simulation can be interesting for vehicular applications, such as the diesel hybrid (battery) vehicle case carried out by Trigui (Trigui et al., 2007).

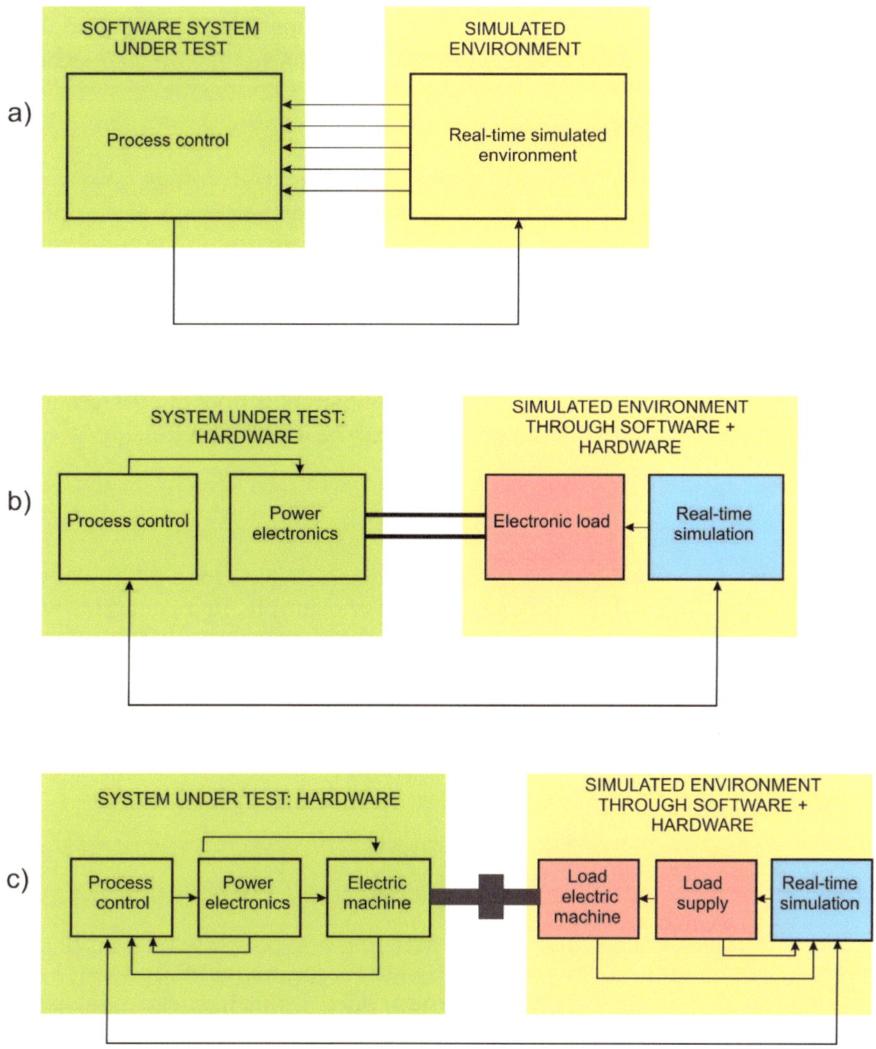


Fig. 1. Hardware-in-the-loop classification in a) signal, b) power and c) mechanical levels

3. **Pure hardware:** In the pure hardware case, the complete hardware setup of hybrid systems presents some drawbacks which cannot be ignored, such as high costs of the elements under test, the infrastructure and security requirements (especially for a hydrogen storage and supply system) and the complexity associated to the performance of the test when a high number of elements are involved. Moreover, each change in the simulation may require a re-design and new setup, with the associated time and cost.

Seen this classification, each HIL type can be applied depending on the final objective of the simulation. The power and mechanical level are particularly useful to simulate energy systems, such as those which can be found in an electric vehicle (batteries, ultracapacitors, fuel cells, etc.). Even if the mechanical HIL simulation can be more easily found in literature (Winkler & Gühmann, 2006) (Timmermans et al., 2007) it presents some drawbacks which can be solved in the HIL power level. These drawbacks are due to the presence of two electric machines, which increases the cost, infrastructure and complexity of the setup. Because the power level avoids the use of electric machines, the approach to the test bench developed can be focused on the energy systems tested, more than on the mechanical part.

## 3. Sizing the HIL simulation platform

To cope with systems where different currents, powers and voltage levels are involved, power engineers make an extensive use of the per-unit system (p.u.). The application of the p.u. system implies the adoption of a set of base values, to which the different magnitudes are referred. Therefore, the p.u. system variables become dimensionless values, allowing a much easier comparison between different alternatives and yielding more meaningful conclusions. The electric variables in a power system are power, voltage, current and impedance and a mathematical relationship between them can be found. As there are four different, but interdependent variables, it is a two degree of freedom system in which once two of the base values are defined, the other two can be obtained with:

$$\begin{aligned} P &= U \cdot I \\ U &= I \cdot Z \end{aligned} \tag{1}$$

The choice of the base values is an arbitrary decision; however, for a particular component it is usual to select its rated values as base values. For example, in a generator with rated 200 MVA and 20 kV voltage, the base power and voltage can coincide with its rated values. Hence, the base current and impedance can be calculated as:

$$I_b = \frac{S_b}{\sqrt{3} \cdot U_b} = \frac{200 \cdot 10^6}{\sqrt{3} \cdot 20 \cdot 10^3} = 577.5\,\text{A} \tag{2}$$

$$Z_b = \frac{U_b^2}{S_b} = \frac{\left(20 \cdot 10^3\right)^2}{200 \cdot 10^6} = 2\,\Omega \tag{3}$$

For this example, if an element consumes 20 MW, the p.u. power will be 0.1 p.u, which gives an understandable measure of the power consumed compared with the rated value.
However, the classical version of the p.u. system does not include the presence of elements with energy storage, as therefore, a base capacity in $A \cdot h$ or C should be taken into account (Gauchia & Sanz, 2008). Hence, it is essential to introduce in the base system the concept of capacity and time, which are related with the discharge duration according to the Peukert equation (Peukert, 1897). $C$ is the battery capacity, $I$ the current and $pc$ the peukert coefficient (usually between 0.5 and 2), which is unique for each technology and model. The equation reveals that the available capacity at constant discharge current is reduced for increasing discharge rates.

$$C \cdot I^{pc-1} = \text{constant} \tag{4}$$

Therefore, a new set of base magnitude which includes a base capacity $C_b$ related to a discharge time $t_b$ must be created. Known the base capacity and the discharge time (information which can be easily found in the data sheet handed by the manufacturer), the base current can be obtained with (5).

$$I_b(A) = \frac{C_b}{t_b} \quad (5)$$

Therefore, the base system for the HIL simulation would include the variables in Table 1.

| Base power (W) | Base voltage (V) | Base current (A) | Base impedance (Ω) | Base capacity (A·h) | Base time (h) |
|---|---|---|---|---|---|
| $P_b$ | $U_b$ | $I_b$ | $Z_b$ | $C_b$ | $t_b$ |

Table 1. Base variables for a HIL simulation which includes energy storage systems.

Any system attributes, as dynamic models, control, setup, results and conclusions expressed in p.u. values are also valid for absolute values, as it does not modify the nonlinear or particular characteristics of the element modelled. In the frame of a HIL simulation, the expression of variables in p.u. can be a very simple and effective way to introduce scaling factors to convert simulated physical variables into full-size physical ones.

## 4. Software and hardware elements of a HIL simulation

A HIL simulation is a flexible way of testing elements which are part of a complex system. Depending on the research interests, a decision must be made about which elements should be simulated and which should be included directly as hardware. When the focus is put directly on the energy systems, it is interesting to include one or more of them as hardware in order to have real information about its operation, whilst the rest of the system can be simulated through software and or hardware elements. When the energy system is the result of an integration of different energy elements, the possibilities of simulated/hardware elements increases significantly.

### 4.1 Case studied

The case studied is depicted in Fig. 2 and represents a hybrid electric vehicle, whose power source is the parallel combination of a fuel cell and battery. The hybrid combination of these two electrochemical energy systems allows to increase the vehicle range and to increase the efficiency, due to the variable driving cycles which can be found. Usually, a vehicle power load profile is formed by a relatively low base power (cruising) with very high abrupt power peak loads (acceleration and overtaking). In vehicles with internal combustion engines, the average power generated by the engine is considerably lower than its maximum power. If the internal combustion engine is substituted by a hybrid system fuel cell/energy storage, the primary system (fuel cell) can have a lower installed power due to the presence of the energy storage.

The fuel cell can supply the power needed for the base power load, however it cannot follow highly dynamic changes in the instantaneous power load. Fast energy storage devices, such as batteries or ultracapacitors can supply the power requested by the peak loads, but they suffer from serious capacity shortage for the supply of the long term energy (Nelson, 2000). As neither the fuel cell nor the fast energy storage devices can supply

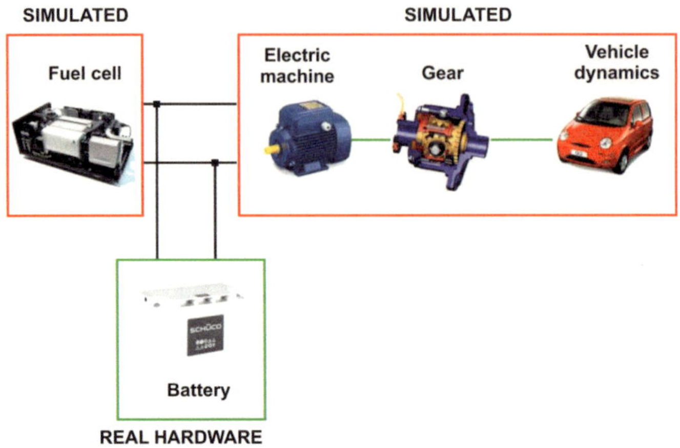


Fig. 2. Case studied: fuel cell-battery hybrid electric vehicle

individually the whole power load, hybridization between both devices is needed. This is, of course, one of the possible alternatives, but it represents the most essential aspects of the problem.

The batteries present a wide variety of commercial technologies; however, especially designed for traction purposes lead-acid batteries are due to their maturity and robustness, as well as to their ability to supply very high current peaks. However, they may be replaced by ultracapacitors or by other battery technologies (Lukic et al., 2008), such as Ni-Mh or Li-ion, the latter being likely the most frequent choice in the next future. Once the elements of the hybrid system have been identified two relevant decisions must be made when designing a HIL simulation. The first one is related to which components should be simulated or implemented by hardware.

This choice is based on the flexibility, cost and complexity which each system adds to the whole HIL simulation. For example, a fuel cell is a complex and high cost system which needs special environments and installations due to the presence of hydrogen. Therefore, this system is a good candidate for being included as simulated in the HIL system. On the other hand, energy storage systems, such as batteries (or supercapacitors) are modular systems which do not require a specific installation and present a lower cost, simple operation and flexibility. Therefore, they are suitable for being part of the hardware systems in the HIL simulation, especially for electrical or electronic research laboratories, which are not specifically designed to operate with hydrogen-fed systems.

## 5. Vehicle modelling

As seen in previous sections, the case studied is a fuel cell/battery vehicle, which is simulated through HIL, with a real battery and simulated fuel cell and vehicle. The simulated fuel cell and vehicle must be done in order to allow a feasible, robust and realistic approach to the real hybrid system. For the purpose of this type of simulation a simplified model for the mechanical part of the vehicle (vehicle forces, gear box and electric machine) may be used. The vehicle power requirements can be calculated through the resistance forces (Gillespie, 1992) which must be overcome by the vehicle power sources to move the vehicle at the desired speed. These resistance forces are normally calculated while the vehicle is being driven uphill, as shown in Fig. 3.

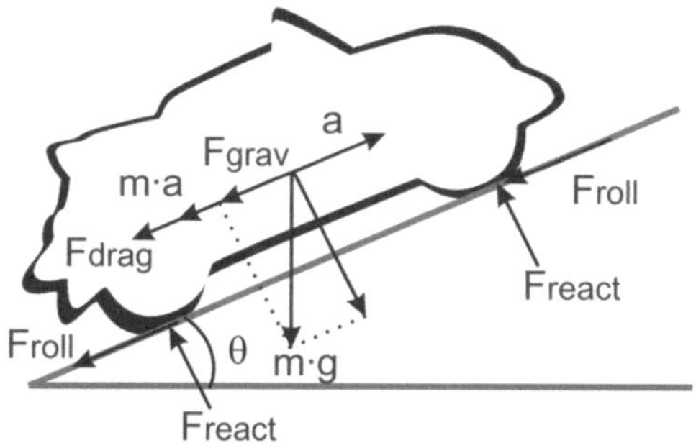


Fig. 3. Resistances to movement present in a vehicle driving uphill.

The resistance forces (6) represent the gravitational force, which depends on the vehicle mass *m*, the gravitational acceleration *g* and the slope angle Θ. $F_a$ represents the acceleration resistance, which depends on the mass *m* and the vehicle acceleration *a*. The drag resistance $F_{drag}$ depends on the air density air, the vehicle frontal surface *A*, the drag coefficient $C_d$ and the vehicle speed *v*. Finally, the rolling resistance $F_{roll}$ depends on the weight, the slope angle *Θ*, the static friction coefficient $f_0$ and the dynamic friction coefficient $f_v$ (see Table 2).

$$
\begin{aligned}
F_{grav} &= m \cdot g \cdot sen\theta \\
F_a &= m \cdot a \\
F_{drag} &= \frac{1}{2} \cdot \rho_{air} \cdot A \cdot C_d \cdot v^2 \\
F_{roll} &= m \cdot g \cdot \cos\theta \cdot \left(f_0 + f_v \cdot v^2\right)
\end{aligned}
\tag{6}
$$

| Parameter | Symbol | Quantity |
|---|---|---|
| Vehicle mass (kg) | *m* | 1000 |
| Frontal surface (m$^2$) | *A* | 1.8 |
| Drag coefficient | $C_d$ | 0.19 |
| Copper losses | $k_c$ | 0.3 |
| Iron losses | $k_i$ | 0.01 |
| Winding losses | $k_w$ | 0.000005 |
| Electronic losses | *EL* | 600 |

Table 2. Vehicle parameters

Electric power trains admit different electric machine technologies, such as induction machine, permanent magnets, switched reluctance, etc. as described by (Ehsani et al., 2007). When the focus is put on the implementation of a flexible test bench which allows both stationary and vehicular applications, the electric machine can be taken into account as a system which introduces an efficiency factor between the power requested by the vehicle and the power supplied by the fuel cell/battery system, centring the attention on the energy sources dynamic behavior and time window. In (7), *T* is the torque needed to move the vehicle at the requested speed; *ω* is the electric machine rotational speed, which depends on the gear ratio, $k_c$, $k_i$ and $k_w$ represent the copper, iron and windage coefficient losses. The values for these coefficients are detailed in Table 2.

### 5.1 Vehicle simulator

The vehicle simulator presented by other authors, such as (Winkler & Gühmann, 2006) or (Timmermans et al., 2007) are constituted by electric machines working against a brake. The inclusion of this electric machine results in a high cost and complex HIL system.

The electric machine and brake can be substituted it by a controlled combination of programmable electronic load and power source. The dc electronic load (Chroma 63201) acts as a sink for the power generated by the energy sources, whilst the dc power source (Sorensen SGI) can generate regenerative braking by injecting power to the dc bus to which the energy sources are connected. It is important to take the regenerative braking into account as it has an important impact on the battery state of charge (*SoC*) and the vehicle range due to the recharging cycles. This vehicle simulator adds flexibility to the whole test bench, as any load power cycle (stationary or vehicular) can be simulated without altering the load simulator.

The layout of the vehicle emulator, shown in Fig. 4, is composed by a software and hardware simulation of the vehicle. The software part includes the driving cycle, vehicle resistive and traction forces, gear ratio and electric machine efficiency and torque. Once the speed driving cycle is transformed to a power or current profile, this profile is used to control externally both the electronic load and power source. The control of both the electronic load and power source is done analogically; through a 0-10 V signal which programs the equipment from 0-100% of its rated value.

The power generated by the simulated fuel cell and the physical battery is absorbed by the electronic load according to the programmed power cycle. The dc power source of the load

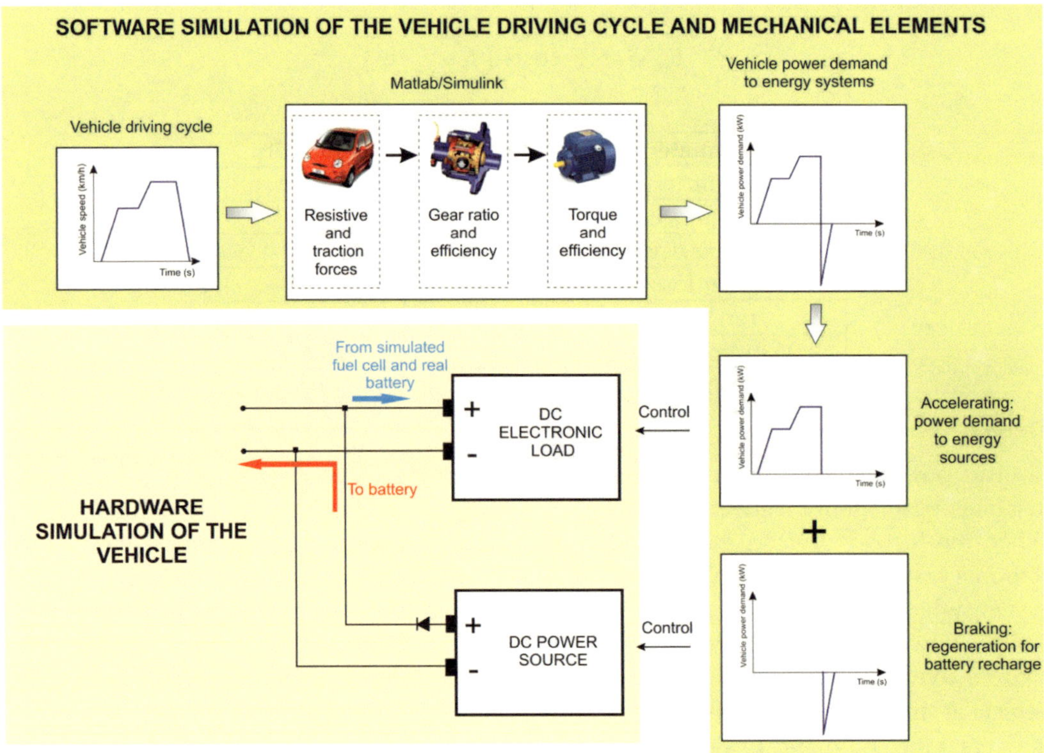


Fig. 4. Vehicle simulator proposed

simulator will only inject power in the dc bus during regenerative braking peaks. This power is used to recharge the battery. The power source is protected with a power diode which prevents current from sinking to the power source. The communication between the software and hardware part of the load simulator is done in real-time, as explained later.

## 6. Fuel cell modelling

This section presents a critical review of the fuel cell modelling techniques proposed by other authors, useful for the testing of hybrid energy systems.
The operational principle of fuel cells is based on electrochemical phenomena, but other processes, such as thermal, chemical, electric, fluid dynamic, etc. are also present. If all these phenomena should be taken into account at a time, the modelling process would result too complex and time consuming. When applying modelling and engineering processes, different approaches can be considered, depending on the final use given to the model developed. For example, fluid dynamic, thermal, chemical and electrochemical approaches are very adequate for the development stage of electrochemical system. But for the application stage, in which the electrochemical system has to interact with the load to which it is connected, an electric modelling of the system could be more adequate. Therefore, in this section we will focus on those models more suitable for a seamless integration into an electric simulation tool.
Engineering models would be useless if the numerical values of their parameters could not be determined through measures. For conventional systems (electric machines, combustion engines, transformers, etc.) whose mathematical models are perfectly defined, there are set of established tests, which allow obtaining the characteristics which define the system under study. Electrochemical systems can be also subjected to a series of tests which allow modelling their electrical behavior. Unlike conventional systems, the structure and mathematical models are still not universally defined, nor the test procedure or parameter obtention univocally established.

### 6.1 Modelling techniques: time vs. frequency domain test

The tests needed to obtain the dynamic models can be carried out either in the time or frequency domain. One of the most extended time domain tests is the current interruption test, whilst the most popular frequency domain test is the electrochemical impedance spectroscopy (EIS). The current interruption test is a time domain test in which the system under study is kept at its operation point (constant current load) until it reaches stationary state. Once reached, the current load is abruptly interrupted, allowing the study of the voltage evolution. Because electrochemical systems operate in direct current (dc), this test is carried out applying a dc current and measuring the dc voltage. The advantage of this modeling technique is its simplicity, both in setup and control. However, there are several drawbacks. One of them is that the model precision depends heavily on the correct identification of the point in which the voltage evolution changes from vertical to nonlinear. An imprecise calculation will cause the incorrect calculation of the voltage drops and the time constant. Finally, this method does not add significant information about the internal processes present in any electrochemical system. Some examples of the application of this method to fuel cells can be found in (Reggiani et al., 2007), and (Adzakpa et al., 2008).
Whilst current interruption is carried out in the time domain and with direct current, EIS is a frequency domain test which needs the application of alternating current and voltage. EIS

tests also seek to calculate the impedance of the system under study. But the most important advantages of frequency domain tests is the richer information obtained and the simpler data processing (if the adequate software is used).

Electrochemical systems present a nonlinear characteristic curve, but can be linearized if small variations are taken into account, as done with small signal analysis. To keep linearity during the tests, the ac signals applied are small enough (e.g. 5% of the rated voltage).

During EIS tests the ripple (either current or voltage) is applied to the electrochemical system. This ripple will cause the system to react, generating an ac voltage (if the excitation signal is current) or ac current (if the excitation signal is voltage). The ripple can be applied with a fixed (not usual) or variable frequency, which in the variable case can be programmed as a sweep. If the imposed ripple is current, it is said to be a galvanostatic mode EIS, whilst if it is a voltage signal it is called a potentiostatic mode. The selection criteria to chose between one mode or another is frequently the control mode of the system under test. For example, the fuel cell current is more easily controlled than the voltage. Hence, it would be easier to apply a galvanostatic (current control) mode.

Electrochemical systems generate direct current, therefore, it is unavoidable to have both dc and ac signals while carrying out EIS tests. The dc level is used to keep the electrochemical system at its operation point, but it is not considered for the impedance calculation, in which only the ac signals are involved. This implies that the dc level must be rejected before the ac impedance is calculated. A diagram explaining the whole process is presented in Fig. 5.

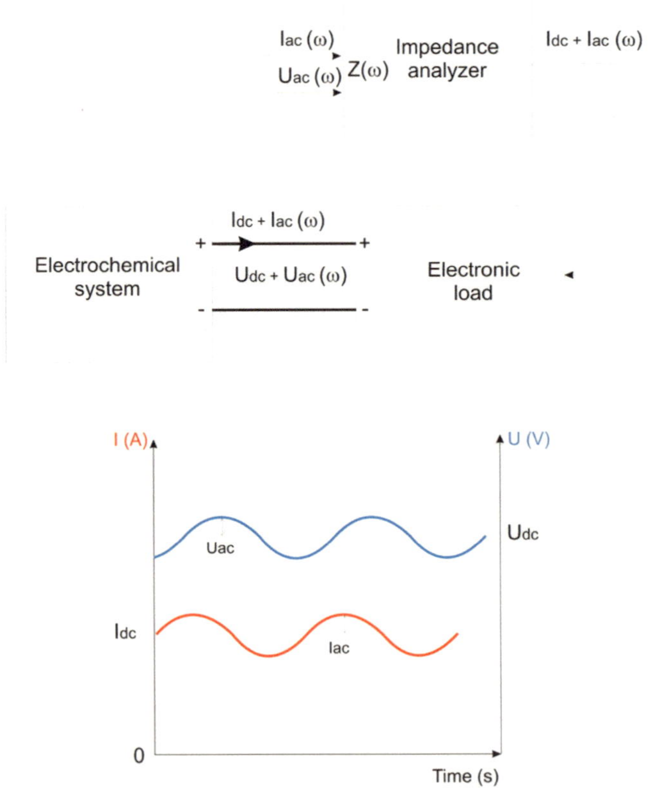


Fig. 5. EIS test procedure

EIS tests can be carried out with off-the-shelf equipment: electronic load, signal generator and voltage and current transducers. However, the subsequent impedance calculation and model extraction is time consuming and complex. Therefore, it is recommendable to use an impedance analyzer, which generates the excitation signal and calculates the complex impedance by measuring the current and voltage.

After the EIS test is carried out, the data must be processed. Normally the impedance analyzer includes a software package to do it. The data are rendered in a text file, which is traduced by the software to a Nyquist and Bode plot. Known these two plots, specially the Nyquist plot, the user can define an equivalent circuit, which the programme fits to the experimental results.

The most frequently used elements are resistances, capacitors and inductances. The resistance is represented by a point on the abscissa axis, with no imaginary part. Ideal capacitances or inductances correspond to vertical lines on the diagram. These ideal elements are rarely, if ever, found. It is more frequent to encounter real systems, which include the association of two or more of these elements, as presented in Fig. 6. The abscissa axis represents the real part of the complex impedance (Z'), whilst the ordinate axis is the imaginary part (Z''), so that the impedance is Z=Z'+jZ''. To facilitate the interpretation of the Nyquist plots, the upper part of the imaginary plot corresponds to the negative imaginary part (-Z'').

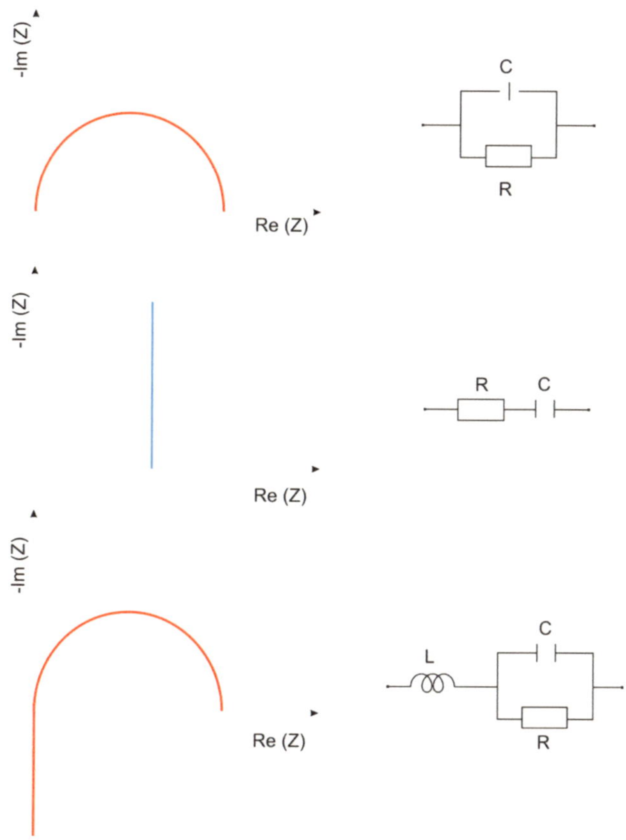


Fig. 6. Nyquist plots for combined ideal elements

For electrochemical systems, classic electric elements (resistances, capacitors and inductances) may not be enough to represent their internal behavior, due to, for example, diffusion phenomena. Most electrochemical systems use porous or rough materials for the electrode manufacture, which affect the diffusion of reactants. As stated by Barsoukov (Barsoukov & Macdonald, 2005), diffusion causes an effect similar to a finite transmission line: the answer of the output to an electric stimulation is delayed, compared to the input. Therefore, the electrochemical system will present a distributed equivalent circuit. The exact impedance cannot be represented as a infinite number of equivalent circuits, so for computational sake, it is normally limited to a finite number.

### 6.3 Fuel cell model

A PEM fuel cell is an electrochemical system which produces the electric power through its principal component: the stack. But the stack needs auxiliary systems, such as fan, compressor, filter, etc. in order to keep the stack environment in the correct operating conditions. Hence, the PEM fuel cell behavior presents thermal effects due to the heat generation, fluid dynamics present by the water and gas transport, electrochemical reactions in the stack, electrical phenomena, etc. The fluid, thermal and electrochemical approaches are specially useful for the development stage of fuel cells. But for the integration in its final application it is more useful an electric model, which can easily interact with the rest of electric models: electric machines, grid, power converters, etc. Therefore, in this chapter, an electric model will be developed. This electric model will adopt the form of an equivalent circuit which will be able to reproduce its voltage dynamic performance. The parameters of the resulting equivalent circuit will be obtained through electrochemical impedance spectroscopy EIS tests. Even if the work is applied to a PEM fuel cell due to its low operation temperature, which is ideal for transportation system, the experimental methodology and results can be extended to other technologies such as SOFC, DMFC, PAFC, etc.

The PEM fuel cell studied in this case is a 1.2 kW Nexa Ballard fuel cell which operates with direct gaseous hydrogen at its anode and air at its cathode.

The frequency electrochemical impedance spectroscopy (EIS) tests were carried out for 10, 20, 30, 40, and 50 A with a frequency interval of 0.5 Hz-6 kHz. The lower limit was chosen at 0.5 Hz as any frequency lower would increase the test time. The higher frequency limit was selected at 6 kHz as higher frequencies would not add relevant information to the model and as the fuel cell, by nature, will never supply high energy peaks during very fast transients. Fig. 7 represents the connections of the experimental test carried out. Throughout the tests the fuel cell is supplied with hydrogen stored at 200 bar. Moreover, the fuel cell communications systems needs a 24 V supply system, therefore a remote controlled power source is used.

The impedance analyzer controls the dc load through its software Zplot™, which establishes the ac and dc current which must be generated by the fuel cell. Both the current and voltage are measured by the impedance analyzer through the terminal V1 and V2 respectively. Later, the software ZView™ processes the experimental results.

The information obtained by ZView™ are the parameters of the equivalent circuit (depicted in Fig. 8) of the PEM fuel cell for the load current tested. Known all these parameters, the relationship of each parameter with the current can be obtained.

Where $E$ represents the open circuit voltage of the fuel cell, which depends on the standard reversible voltage $E_0$, the partial pressures of hydrogen $p_{H2}$, oxygen $p_{O2}$, and water $p_{H20}$, as well as on the number of cells $N$, the ideal gas constant $R$, the temperature $T$, and the

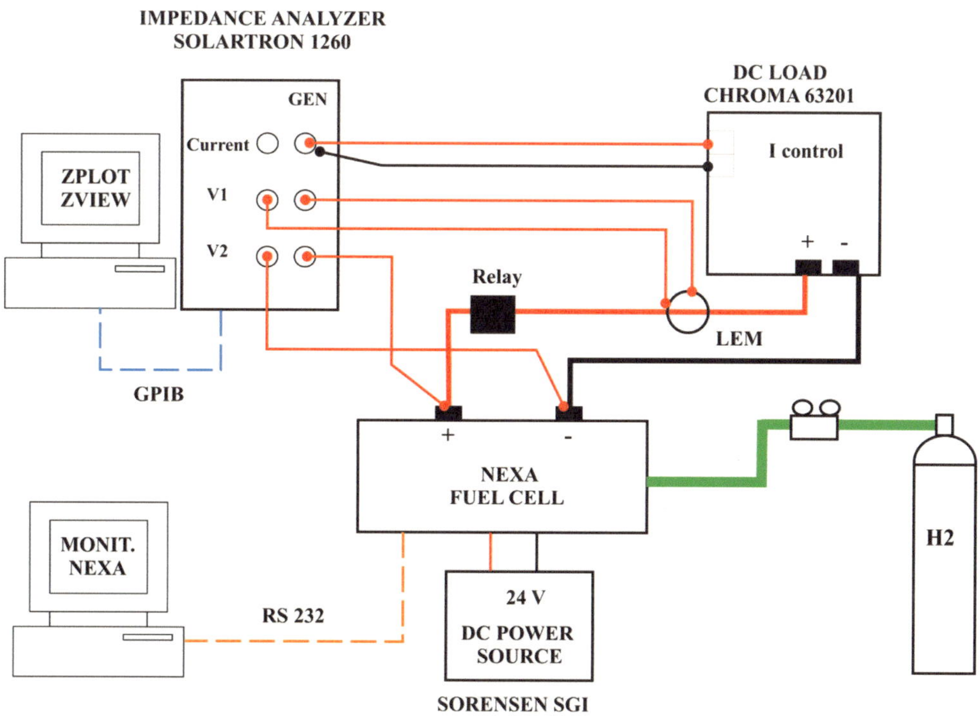


Fig. 7. EIS experimental test setup

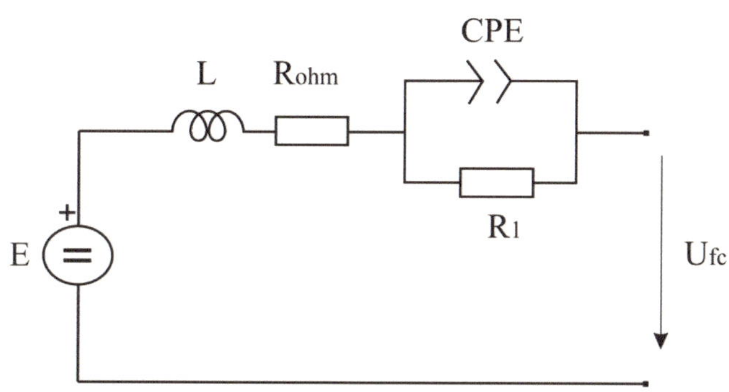


Fig. 8. Fuel cell equivalent circuit

Faraday constant *F*. The series resistance $R_{ohm}$ reflects the sum of the electrode internal resistance and the resistance to the flow of protons inside the membrane. *L* represents the inductive behavior of the fuel cell at high frequencies and the network $R_1CPE$ represent the dynamic behavior of the double layer capacitor which is situated between each electrode and the electrolyte. A constant phase element (*CPE*) is an element which represents the electrode roughness and inhomogeneous distribution of reactives on the electrode surface. As a *CPE* is not a physical electric element (Kötz & Carlen, 2000), it can be approximated by an equivalent capacitor $C_1$. Once known the equivalent circuit of the fuel cell, the fuel cell output voltage can be calculated with:

$$U_{fc} = E - R_{ohm} \cdot I - \left( I - C_1 \frac{dU_{C_1}}{dt} \right) \cdot R_1$$
$$E = N \cdot \left[ E_0 + \frac{R \cdot T}{2 \cdot F} \cdot \left( \ln \frac{p_{H_2} \cdot \sqrt{p_{O_2}}}{p_{H_2O}} \right) \right] \quad (7)$$

A Matlab™/Simulink™ model of the fuel cell was programmed. In Fig. 9 experimental and model results are compared when a dynamic current is requested to the fuel cell. It can be observed that the model follows adequately the real fuel cell dynamic behavior. The discrepancy between the model and experimental results is due to the differences of the real and simulated temperature during very high currents. When the fuel cell stack reaches 65 ºC the cooling fan increases its duty cycle in order to cool the stack.

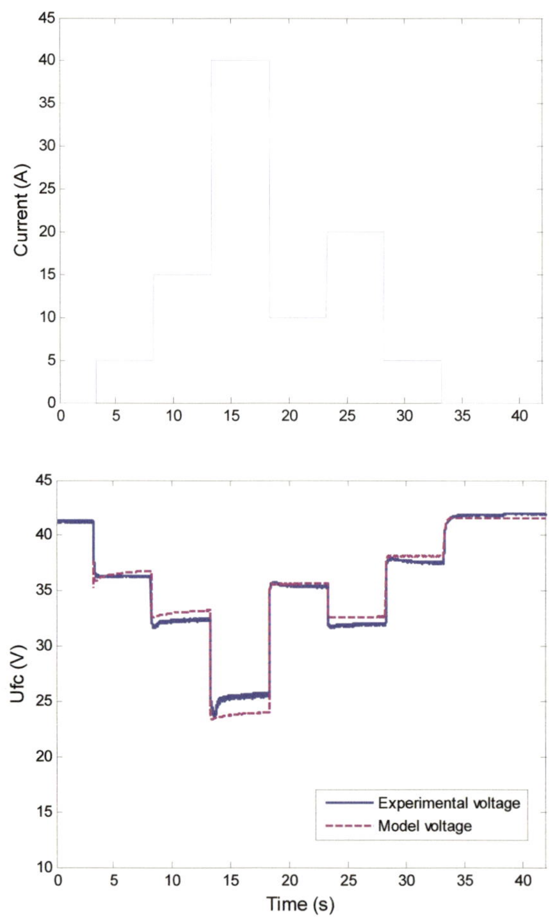


Fig. 9. Model validation

### 6.4 Fuel cell simulator

The fuel cell simulator is based on the dynamic nonlinear model obtained previously, which is fed to a dc power source which emulates its behavior. The fuel cell system simulator is formed by a programmable dc power source (Sorensen DCS20-150E) which is programmed with a 0-10 V signal for the output voltage and with a 0-10 V signal for the output current limit, allowing the voltage and current control of the fuel cell, just as it happens when the fuel cell is connected to a dc-dc power converter. These values are established by the control system, depending on the vehicle driving cycle and the battery state of charge, as explained later. The model presented in the previous subsection is implemented in Matlab™/Simulink™ and fed to the dc power source, as shown in Fig. 10. Just as in a physical fuel cell system, the power source which simulates the fuel cell operation is protected against sinking currents by a power diode. As in the case of the vehicle simulator, the interaction between the software and hardware simulation of the fuel cell system is done in real-time.

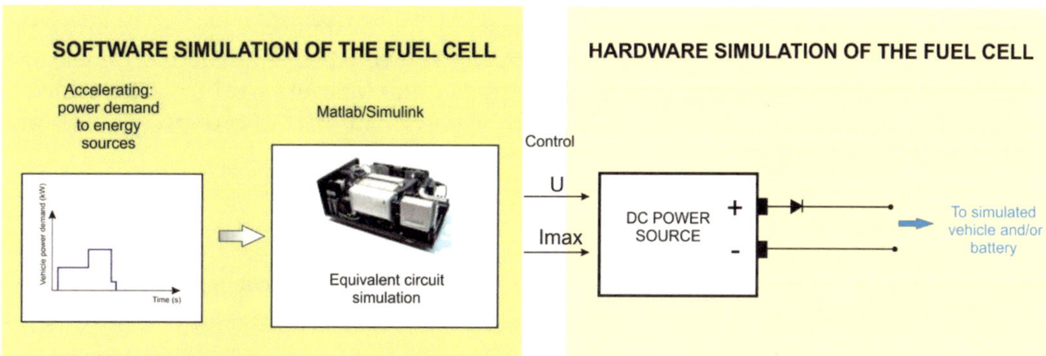


Fig. 10. Fuel cell simulator

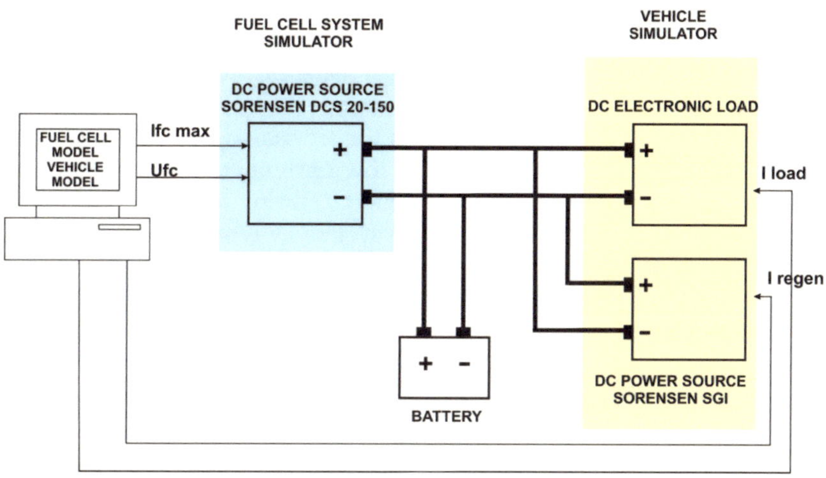


Fig. 11. HIL test bench power connections

## 7. HIL test bench

### 7.1 Power connections

The energy storage subsystem of this hybrid energy application consists of a Maxxima Exide-Tudor sealed lead-acid battery of 12 V, 50 Ah (20 h) connected in parallel to the dc bus. The power connections for the whole test bench are depicted in Fig. 11. As it can be observed, multiple power sources are connected in parallel. However, this presents no inconvenience in terms of voltage or current establishment. This is due to two facts. The first one is that the vehicle simulator (dc electronic load + power source) is current controlled, and therefore adapts to the voltage imposed by the hybrid energy system and hence, no ideal voltage sources are directly connected in parallel. The power source which simulates the fuel cell acts as a voltage or current source, depending on the load power requirements. Besides, both the simulated fuel cell and the battery show a non-negligible variable series impedances.

### 7.2 HIL acquisition system

An acquisition system is necessary to register the evolution of the different devices involved. LEM™ transducers are used to measure the dc bus voltage, as well as the simulated fuel cell, dc electronic load and dc power source which emulate the vehicle and battery current. The signals are acquired through an input/output board DS 2201, which is connected to its connector panel CP 2201. Both elements are part of a dSpace™ real-time control and acquisition system.

### 7.3 HIL control system

Hybrid systems accept a variety of control schemes, as for example (Thounthong et al., 2006), (Jiang et al., 2005), (Jiang & Fahini, 2009) or (Gao & Ehsani, 2009). The objective of the control system presented is to test the HIL test bench itself, in order to demonstrate its feasibility. Due to the fact that the energy system is hybrid between a simulated fuel cell and a physical battery, the control applied in this case will analyze and quantify the power share between the simulated fuel cell system and the real battery. Because the fuel cell system is not able to supply high energy peaks, the fuel cell system will be assigned the role of supplying the average power. With this strategy, the simulated fuel cell can be kept at or near the most efficient operation point. The power peaks will be supplied by the battery, which has a much faster response. The result is the simulated fuel cell feeding the average load power and recharging the battery during low current consumption periods, and the battery supplying power peaks and accepting energy during regenerative braking. Thence, it is possible to have a total control of the hybrid system by what could be called the "energy management mode" (EMM) of the system, which will depend on the type of control selected: fuel cell optimization, battery *SoC*, etc. The battery *SoC* at any instant must be calculated (8) and controlled.

$$SoC_t(\%) = SoC_{t-1} \pm \frac{\eta}{C_n} \cdot \int I(t) \cdot dt \cdot 100 \qquad (8)$$

The *SoC* calculation takes into account the charging efficiency and the battery rated capacity $C_n$. As reflected in the second addend of (8), the charged capacity depends on the charge effciency. The sign of the second addend depends on the battery operation mode: it is positive for charge and negative for discharge.
The complete Matlab™/Simulink™ programme is shown in Fig. 12. The control system is implemented in a real-time platform as a general-purpose operating system does not have a consistent, repeatable, and known timing performance. The real-time platform is based on a dSpace™ Modular Hardware formed by an expansion box PX10, with a digital signal processor DSP DS 1006. The DSP communicates with the host computer through an optical

cable and a PCI board. The dSpace and the host computer carry out task sharing, as the dSpace hardware performs the real-time calculation and the host computer provides the user interface and the experiment environment. The DSP DS 1006 is built around an AMD Opteron processor which runs at 3 GHz and has 256 MB as local memory for the execution of real-time models, 128 MB of global memory to exchange data with the host computer and 2 MB of flash memory. The I/O boards can be programmed using Matlab™/Simulink™ and Real Time Interface (RTI). The program developed is executed with a fixed step 1 ms and a fourth order Runge Kutta algorithm. The 1 ms time step selected is smaller than the vehicle (seconds), fuel cell (20 ms for the model used) and battery (10 ms for the battery used) time constant, and can therefore capture the dynamic operation of any system involved. Smaller time steps could lead to heavy experimental data files (e.g. over 1 million samples were recorded for each signal captured during the test carried out.) Fig. 13 presents the layout and photograph of the whole HIL simulation.

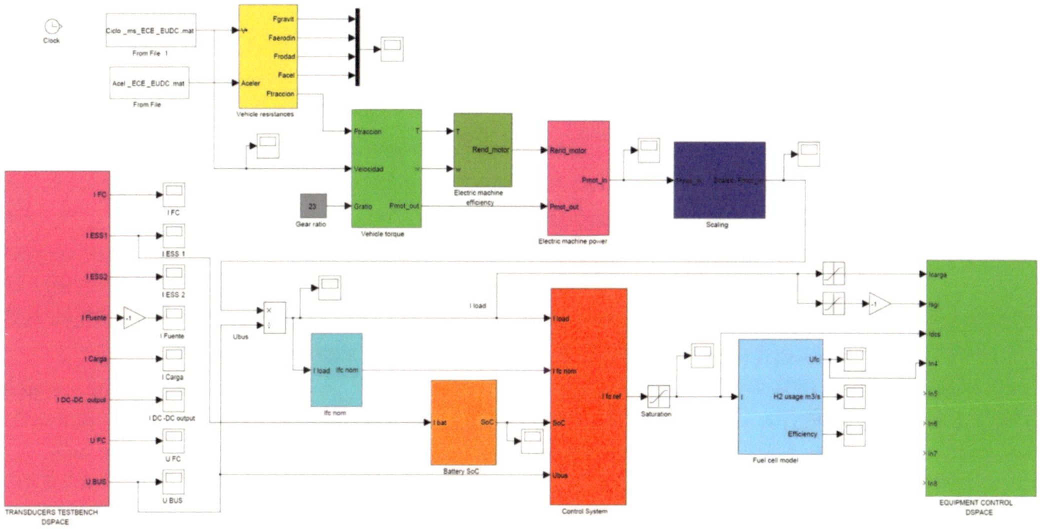


Fig. 12. HIL Matlab™/Simulink™ programme

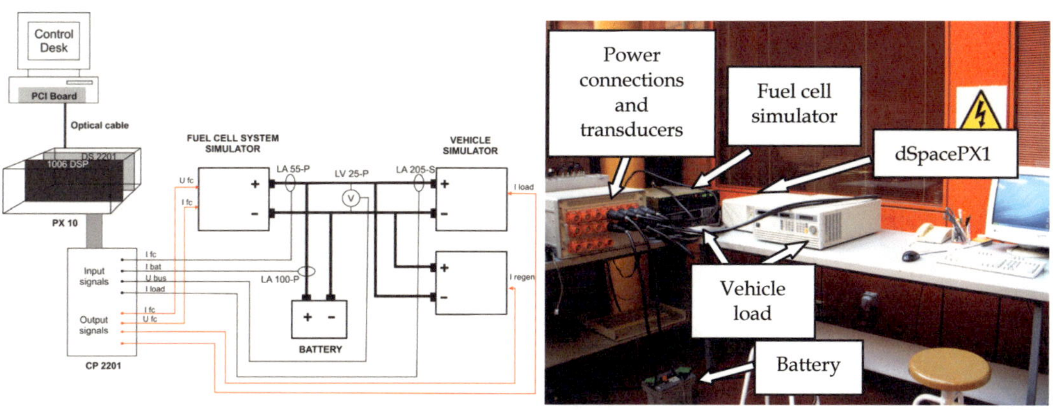


Fig. 13. Complete HIL setup and photograph

### 7.4 Energy management modes EMM's

In a hybrid system, a wide range of control schemes could be implemented, depending on the objective: maximum range, minimum fuel consumption, minimum *SoC* variation, etc. This wide range of control schemes is due to the hybrid nature of the system, with one source able of supplying long term load (fuel cell) and another source capable of supplying shorter term loads (energy storage system). With these different time windows, control schemes may benefit one system but penalize the other. Therefore, some type of compromise should be reached.

In this work, two different and representative EMM's of how a control scheme affects each system will be applied: the first one will reduce the hydrogen consumption and efficiently recharge the battery, whilst the second one will keep the battery *SoC* within an established interval. Of course, other control schemes could be tested.

For $EMM_1$, the fuel cell output will vary between two different levels (one low level to reduce the fuel consumption and the maximum power point when the load current surpasses 100 A) while supplying the load and recharging the battery with a profile where the values are selected based on previous experiences with the battery. If the load current is smaller than the rated fuel cell current the battery starts a recharge cycle. The charge current depends on the battery *SoC*, as for increasing *SoCs* the battery voltage increases. The recommended charging voltage limit for this battery, as specified by the manufacturer, is 14.4 V. Due to the fact that the charge effciency decreases dramatically for high *SoCs*, the *SoC* is kept between 40 %-60 %, as in this interval there is no risk of overcharge or battery depletion, which would contribute to shorten the battery life due to gasification or sulphatation phenomena. The 60% upper limit for the *SoC* also assures that the battery will be always ready to accept power peaks during regenerative breaking.

For $EMM_2$, the fuel cell will be kept at its maximum power point, both during the vehicle power following and during the battery recharge. Hence, the battery will have to deal with the peak transients. If during the driving cycle the battery voltage surpasses the manufacturer recommendation (max. 14.4 V or 1.2 p.u.) the control system will reduce the fuel cell reference (37.78 A or 2 p.u.) in order to reduce the charge regime or force the battery into a discharge cycle, which in both cases will lower the battery voltage.

In general, for both EMM's, depending on the drive cycle, the battery could exceed the maximum 100% *SoC* if a low load or high regenerative breaking takes place. Hence, if the maximum *SoC* is exceeded the fuel cell reference will be reduced in order to force the battery into a discharge cycle. Also, the fuel cell will be in continuous operation, even during short stops, in order to recharge the battery. In this particular case, if the stop time exceeds 5 minutes the control system will stop the fuel cell. Moreover, if during the stop time the battery *SoC* exceeds the 80 %, the fuel cell will also stop.

## 8. HIL results

To test the p.u. HIL simulation presented, a driving cycle has been applied to the simulated system. The New European Driving Cycle (NEDC) simulates during 1225 seconds an urban and suburban route with frequent stops, as it can be seen in Fig. 14. The maximum speed is 120 km/h.

With this driving cycle the power requested to the downsized fuel cell/battery system is shown in Fig. 15. The maximum downsized power is 2500 W, which corresponds to a real 25

kW of the original application. As the simulation is carried out for a smaller system (10:1), the results presented are expressed in per-unit values.
The simulated fuel cell current is measured by a current transducer at the Sorensen DCS power source which emulates its behavior for both control schemes. The current profile is presented in per-unit values. For $EMM_1$ in Fig. 16 the fuel cell current varies between 1.2 p.u. and 2 p.u., which are the two levels established. There is also a third level at 0.5 p.u., which corresponds to the battery recharge. For $EMM_2$ in Fig. 17 the fuel cell is kept at a constant operation point (its maximum power transfer point), which corresponds with a 2 p.u. current, when the current is referred to the base system. These current values affect the hydrogen consumption, which is 2.5e-3 l/s for $EMM_1$ and 3.7e-3 l/s for $EMM_2$. Due to the fact that this constant operation of the fuel cell will affect the battery *SoC* and voltage, the control system will measure continuously the battery voltage. It can be observed in Fig. 16 that when the measured battery voltage surpasses 14.4 V (1.2 p.u.) the fuel cell current decreases in order to allow a battery discharge or reduce the recharge level. On the other hand, the bus voltage variation for control scheme 1 is lower, as it varies between 0.87 and 1.2 p.u.

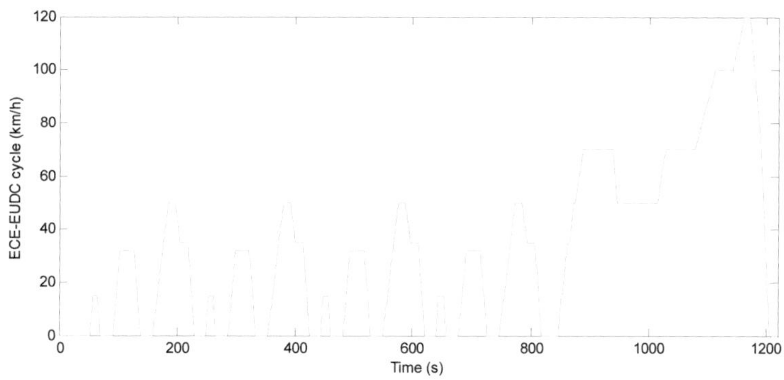


Fig. 14. Driving cycle

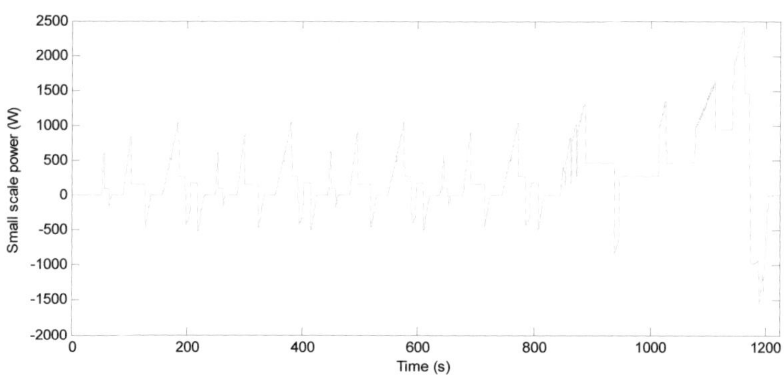


Fig. 15. Downsized power cycle (10:1)

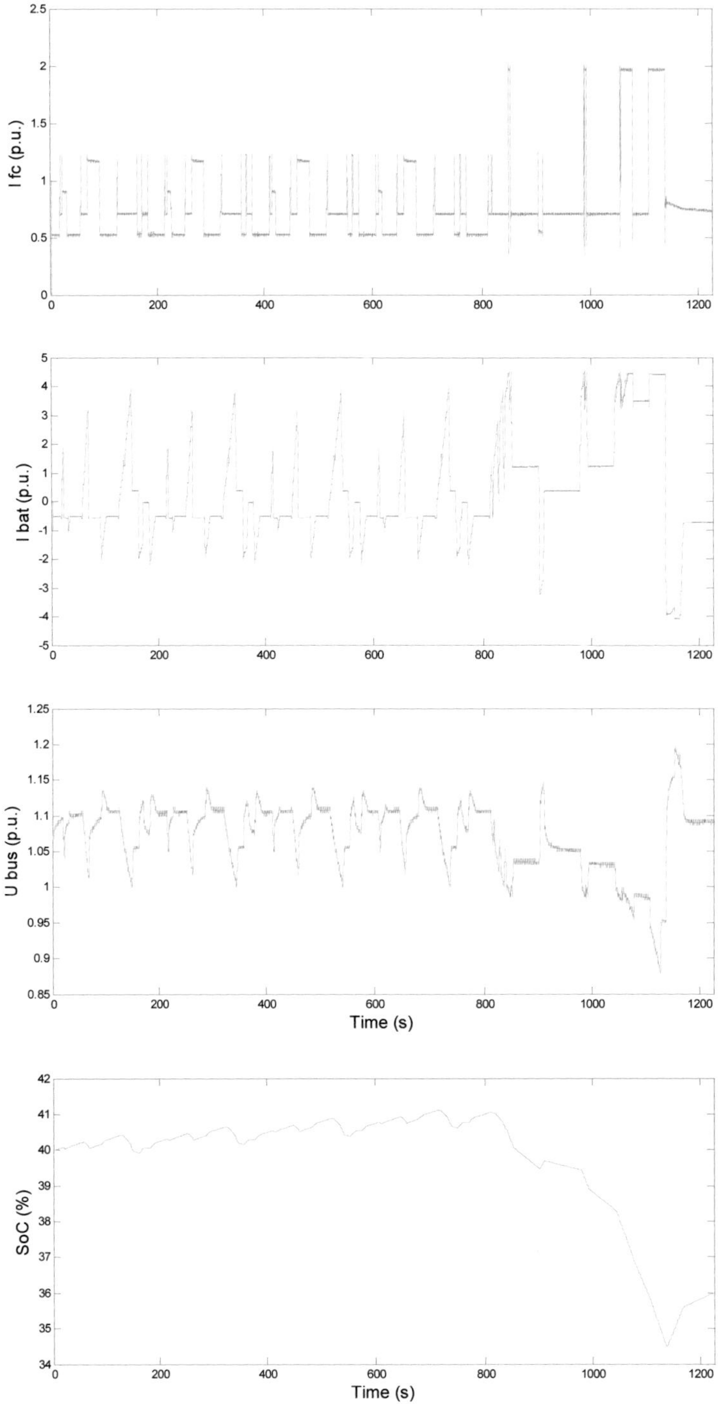


Fig. 16. Energy management mode 1

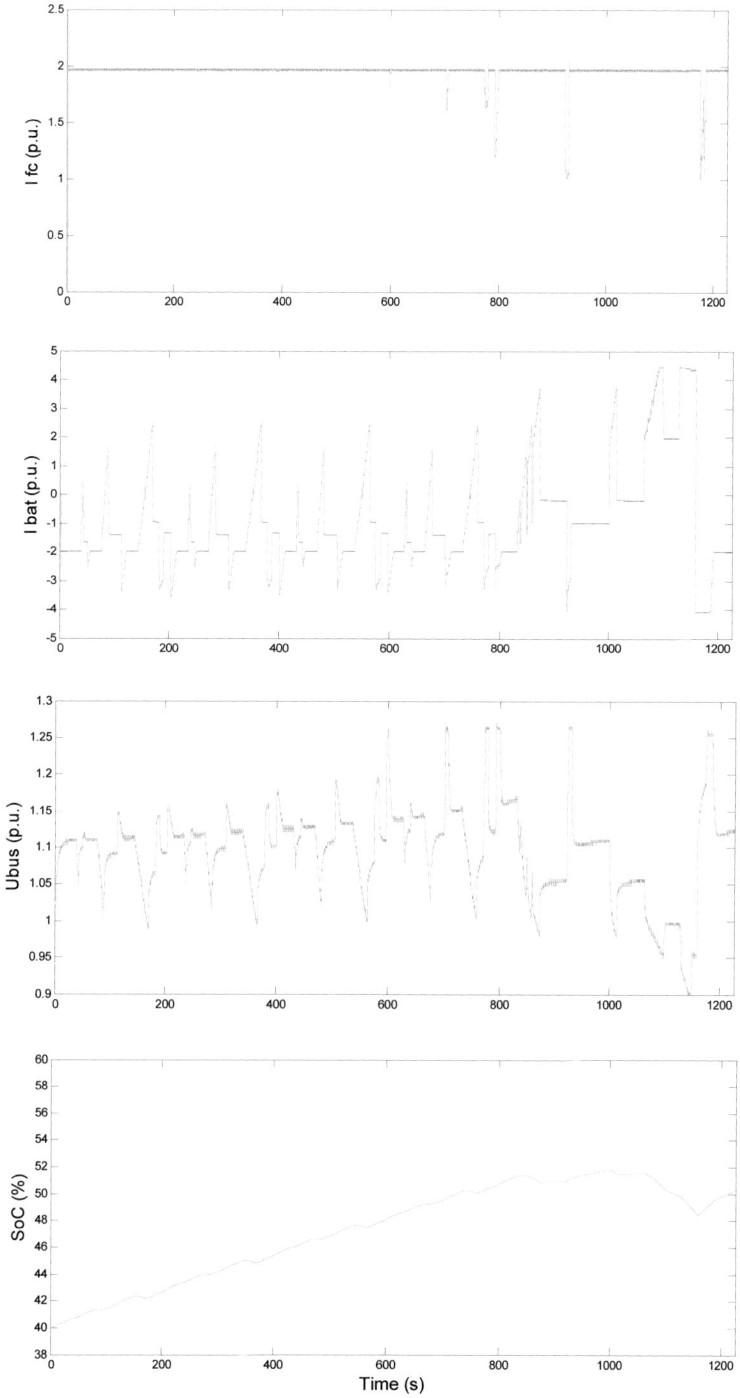


Fig. 17. Energy management mode 2

For both energy management modes the battery absorbs up to 5 p.u. The battery is recharged both from the fuel cell during low loads or during the regenerative braking. As the battery is reserved for the high peak currents, the bus voltage presents a variation around 1.07 p.u. for $EMM_1$ and around 1.1 p.u. for $EMM_2$. The voltage variation is of 0.33 p.u. for $EMM_1$ and 0.37 p.u. for $EMM_2$. This voltage variation is a normal situation for a battery, which needs to vary considerably its voltage in order to supply the current demanded during the charge/discharge cycles. This voltage variation is not so dramatic in other energy storage systems. For example, supercapacitors can supply or absorb hundreds of amperes during very short time intervals by just varying mV the supercapacitor voltage. The battery absorbs the frequent regenerative braking and fuel cell current which keeps the battery *SoC*. However, the two different energy managements affect the *SoC* in a considerable way. For $EMM_1$ the *SoC* does not vary significantly during most part of the cycle, but collapses during the suburban section. $EMM_2$ keeps the battery *SoC* within more appropriate values: 40% (0.4 p.u.) and 52% (0.52 p.u.), which avoids the battery depletion or overcharge. The fuel cell power varies around 2.2 p.u., which corresponds to 550 W in the downsized system and 5.5 kW in the real application. The battery power can reach nearly 5 p.u., which is 1250 W in the downsized system and 12.5 kW in the real system.

## 9. Conclusion

Hardware-in-the-loop simulation is a powerful tool for simulating systems with a high number of components, which may require a complex and expensive setup. In this chapter a HIL simulation has been applied to a fuel cell/battery hybrid vehicle. Unlike other authors, the vehicle simulation does not include an electric machine to reproduce the regenerative braking. A combined control of a dc electronic load and dc power source allows simulating both the vehicle power requirement and regenerative braking. This vehicle simulator can easily switch to a stationary load simulator by just changing the programmed power cycle. The hybrid fuel cell/battery system is setup as a combination of simulated and real hardware systems. The fuel cell simulator can be setup with a programmed dc power source which is able to reproduce the fuel cell voltage and current evolution. On the other hand, the battery is a simple and modular system which can be easily introduced as hardware.
The HIL simulation is carried out under a p.u. system, which allows to downsize the whole test bench and to study the hybridization between simulated fuel cell and battery. Moreover, different control strategies or EMM's can be tested to simulate different scenarios.

## 10. References

Maclay, D. (1997). Simulation gets in the loop, *IEE Review*, vol. 43, pp. 109-112.

Bouscayrol, A. (2008). Different types of hardware-in-the-loop simulation for electric drives, *IEEE International Symposium on Industrial Electronics*, Cambridge (U.K.).

Lu, B.; Wu, X.; Figueroa, H. & Monti, A. (2007). A low cost real time hardware-in-the-loop testing approach of power electronics controls, *IEEE Transactions on Industrial Electronics*, vol. 54, pp. 919-931.

Ren, W.; Steurer, M. & Qi, L. (2008). Evaluating dynamic performance of modern electric drives via power hardware-in-the-loop simulation, *IEEE International Symposium on Industrial Electronics*, Cambridge (U.K.).

Trigui, R; Jeanneret, B.; Malaquin, B.; Badin, F. & Plasse, C. (2007). Hardware in the loop simulation of a diesel parallel mild-hybrid electric vehicle, *IEEE Vehicle Power Propulsion Conference*, pp. 448-455.

Winkler, D. & Gühmann, C. (2006). Hardware-in-the-loop simulation of a hybrid electric vehicle using Modelica/Dymola, *22nd International Battery, Hybrid and Fuel Cell Electric Vehicle Symposium and Exhibition*, Yokohama (Japan), pp. 1054-1063.

Timmermans, J.M.; Van Mierlo, J. & Lataire, P. (2007). Test platform for hybrid electric power systems: development of a HIL test platform, *European Conference on Power Electronics and Applications*, Aalborg (Denmark), pp 1-7.

Gauchia, L. & Sanz J. (2009). Per unit representation of electrical magnitudes in batteries: a tool for comparison and design, *Energy Conversion and Management*, vol. 50, pp. 554-560.

Peukert, W. (1897). Ueber die Abhängigkeit der kKapacität von der Entladestromstärke bei Bleiakkumulatoren, *Elektrotechnische Zeitschrift*, vol. 20, pp. 287-288.

Nelson, R. (2000). Power requirements for batteries in hybrid electric vehicles, *Journal of Power Sources*, vol. 157, pp. 2-26.

Lukic, S.M.; Ciao, J.; Bansal, R.C.; Rodriguez, F. & Emadi, A. (2008). Energy storage systems for automotive applications, *IEEE Transactions on Industrial Electronics*, vol. 55, pp. 2258-2267.

Gillespie, T. (1992). Fundamentals of Vehicle Dynamics. *SAE*.

Ehsani, M.; Gao, Y & Miller, J. (2007). Hybrid electric vehicles: architecture and motor drives, *Proceedings of the IEEE*, vol. 95, pp. 719-728.

Reggiani, U.; Sandrolini, L.; & Burbui, G. (2007). Modelling a PEM fuel cell stack with a nonlinear equivalent circuit, *Journal of Power Sources*, vol. 165, pp. 224-231.

Adzakpa, K.; Agbossou, Y.; Dube, Y.; Dostie, M.; Fournier, M. & Poulin, A. (2008). PEM fuel cells modeling and analysis through current and voltage transient behaviors, *IEEE Transaction on Energy Conversion*, vol. 23, pp. 581-591.

Barsoukov, E. & Macdonald, J. (2005). Impedance spectroscopy, Wiley & Sons.

Kötz, R. & Carlen, M. (2000). Principles and applications of electrochemical capacitors, *Electrochmica Acta*, vol, 45, pp. 2483-2498.

Thounthong, P.; Raël, S.: Davat, B & Sadli, I. (2006). A control strategy of fuel cell/battery hybrid power source for electric vehicle applications, *37th IEEE Power Electronic Specialists Conference*, pp. 1-7, Jeju (South Korea).

Jiang, W.; Gao, L. & R.A. Dougal. (2005). A flexible multiobjective control in active hybrid fuel cell-battery hybrid power system, *IEEE Transactions on Power Electronics*, vol. 20, pp. 244-253.

Jiang, W. & Fahini, B. (2009). Active current sharing and source management in a fuel cell-battery hybrid power system, *IEEE Transactions on Industrial Electronics*, vol. 56, pp. 1-1.

Gao, Y. & Ehsani, M. (2009). Design and control methodology of plug-in hybrid electric vehicles, *IEEE Transactions on Industrial Electronics*, vol. 56, pp. 1-1.

# 8

# Analysis of the Regenerative Braking System for a Hybrid Electric Vehicle using Electro-Mechanical Brakes

Ki Hwa Jung, Donghyun Kim, Hyunsoo Kim and Sung-Ho Hwang
*Sungkyunkwan University*
*Republic of Korea*

## 1. Introduction

In the future, automobile makers will be required to produce new technologies that reduce automotive emissions while still satisfying the ever increasing performance demand of drivers. Active safety control systems such as Anti-lock Brake System (ABS), Electronic Braking force Distribution (EBD), Traction Control System (TCS) and Electronic Stability Program (ESP) need to improve their existing braking functions in order to be truly effective in improving driving safety. Therefore, brake systems will need to be faster and more sophisticated when controlling braking forces at the wheels. In addition, smaller pedal pressure and reduced stroke will be required to produce a larger braking force. With ABSs, the surge and fluctuation of pedal force gives the driver an uncomfortable feeling. These are only a few of the problems and technical limitations of current braking control systems (Semm et al., 2003, Peng et al., 2008).
Figure 1 shows the development trend of braking control systems. The future development in braking technology will progress towards brake-by-wire; therefore, brake manufacturers will need to take a greater interest in the development of Electro-Mechanical Brake (EMB) systems (Line et al., 2004, Emereole & Good, 2005)
EMB systems replace conventional hydraulic braking systems by eliminating the hydraulics and replacing them with electrical components. They are able to eliminate the large vacuum booster found in conventional systems, which helps to simplify production of right- and left-hand drive vehicle variants. When compared to conventional braking systems, EMB systems offer increased flexibility for components placement by totally eliminating the hydraulic system (Nakamura et al., 2002). Figure 2 shows the comparison of EMB and EHB (Electro-Hydraulic Brake) systems.
This paper investigates the modeling and simulation of EMB systems for Hybrid Electric Vehicles (HEV). The HEV powertrain was modeled to include the internal combustion engine, electric motor, battery, and transmission. The performance simulation for the regenerative braking system of the HEV was performed using MATLAB/Simulink. The control performance of the EMB system was evaluated via simulation of the regenerative braking of the HEV during various driving conditions.

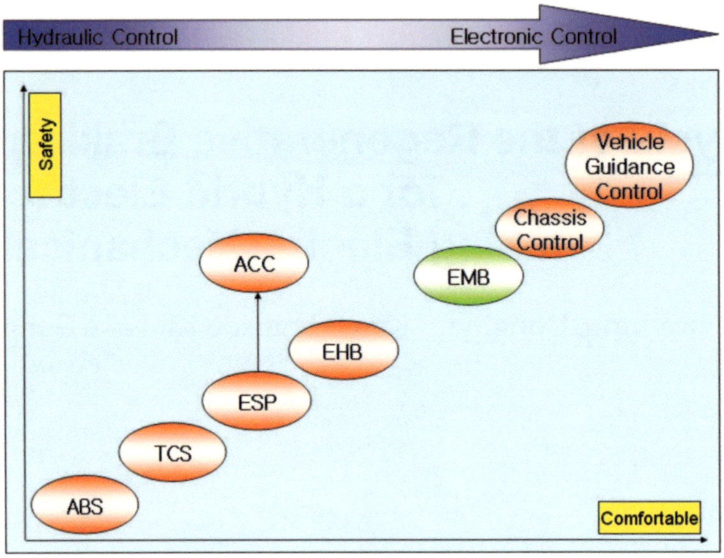


Fig. 1. Development trend of brake control systems

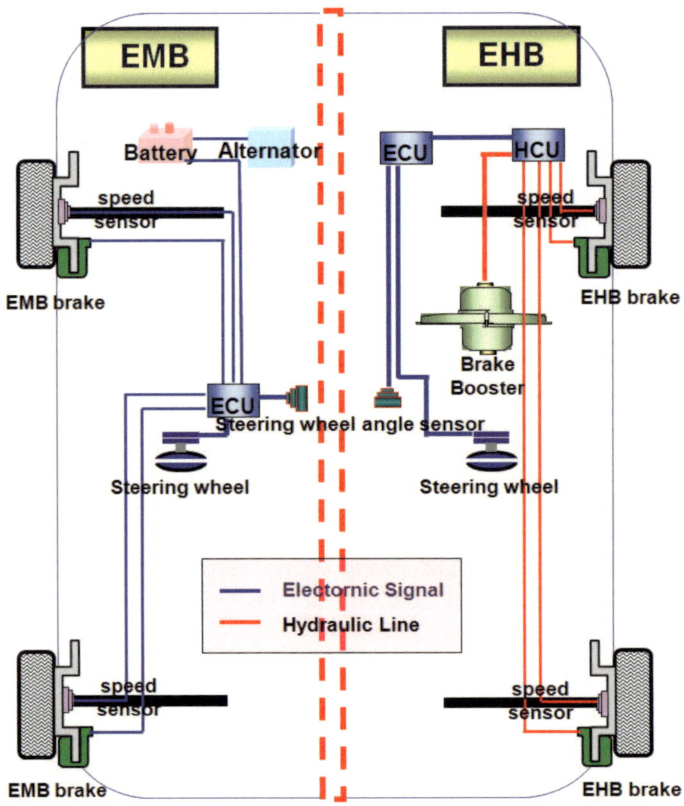


Fig. 2. Comparison of EMB and EHB systems

## 2. HEV powertrain modeling

Figure 3 shows the structure of the HEV investigated in this paper. The power source of this HEV is a 1.4 liter internal combustion engine and a 24 kW electric motor connected to one of the axes. The transmission and braking system are an Automated Manual Transmission (AMT) and an EMB system with pedal stroke simulator, respectively. EMB supplies braking torque to all four wheels independently, and the pedal stroke simulator mimics the feeling of the brake pedal on the driver's foot.

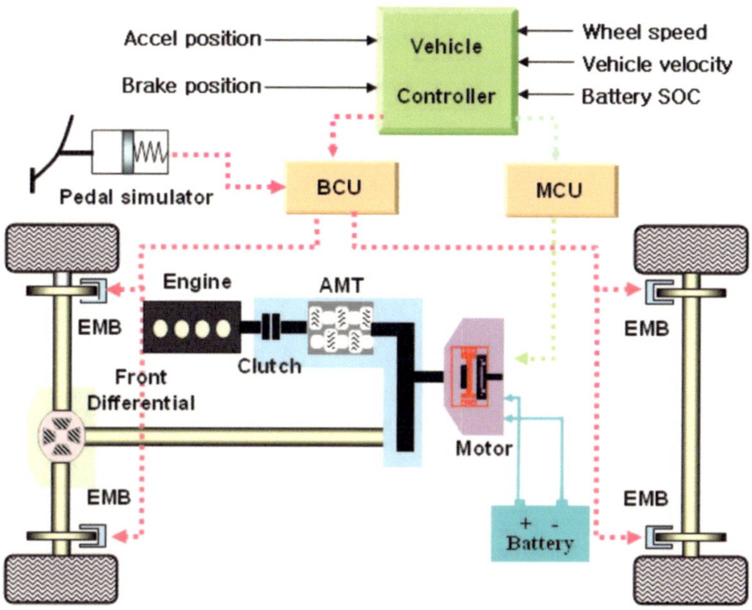


Fig. 3. Configuration of HEV braking control system

The vehicle controller determines the regenerative braking torque and the EMB torque according to various driving conditions such as driver input, vehicle velocity, battery State of Charge (SOC), and motor characteristics. The Motor Control Unit (MCU) controls the regenerative braking torque through command signals from the vehicle controller. The Brake Control Unit (BCU) receives input from the driver via an electronic pedal and stroke simulator, then transmits the braking command signals to each EMB. This is determined by the regenerative braking control algorithm from the value of remaining braking torque minus the regenerative braking torque. The braking friction torque is generated when the EMB in each wheel creates a suitable braking torque for the motor; the torque is then transmitted through the gear mechanism to the caliper (Ahn et al., 2009).

### 2.1 Engine

Figure 4 shows the engine characteristic map used in this paper. The complicated characteristics of this engine are due to many factors, such as fuel injection time, ignition time, and combustion process. This study uses an approximated model along with the steady state characteristic curve shown in Figure 4.
The dynamics of the engine can be expressed in the following equation:

$$J_e\dot{\omega}_e = T_e(\theta,\omega_e) - T_{loss} - T_{clutch} \quad (1)$$

where $J_e$ is the rotational inertia, $\omega_e$ is the engine rpm, $T_e$ is the engine torque, $T_{loss}$ is loss in engine torque, and $T_{clutch}$ is the clutch torque.

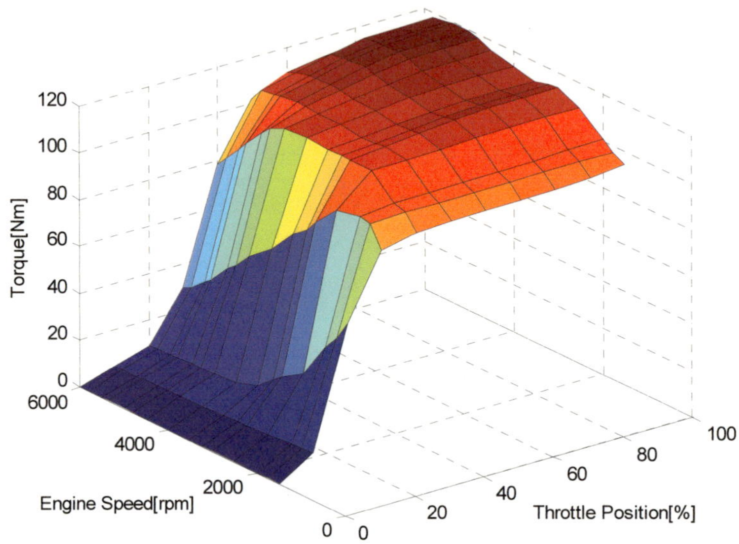


Fig. 4. Engine characteristic map

**2.2 Motor**

Figure 5 shows the characteristic curve of the 24 kW BLDC motor used in this study. In driving mode, the motor is used as an actuator; however, in the regenerative braking mode, it functions as a generator.

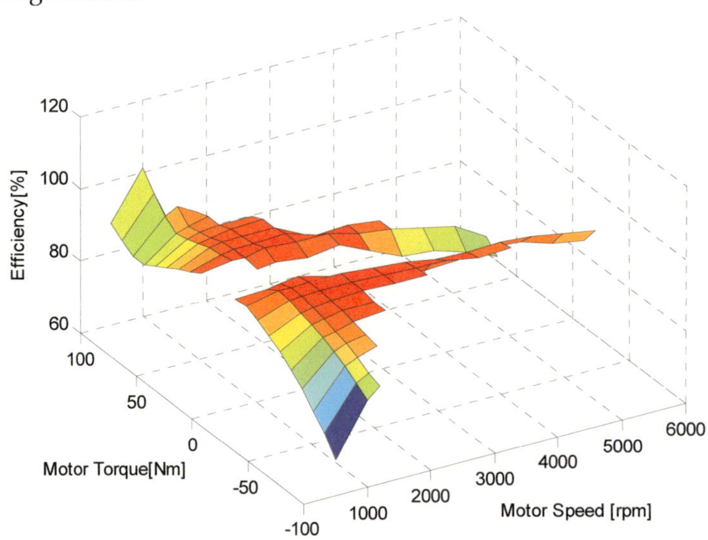


Fig. 5. Characteristic map of the motor

When the motor is functioning as an actuator, the torque can be approximated using the following 1st order equation:

$$\frac{dT_m}{dt} = \frac{T_{m\_desired} - T_m}{\tau_{T_m}} \tag{2}$$

where $T_m$ is the motor torque, $T_{m\_desired}$ is the required torque, and $\tau_{T_m}$ is the time constant for the motor.

### 2.3 Battery

The battery should take into account the relationship between the State Of Charge (SOC) and its charging characteristics. In this paper, the input/output power and SOC of the battery are calculated using the internal resistance model of the battery. The internal resistance is obtained through experiments on the SOC of the battery. The following equations describe the battery's SOC at discharge and charge.

- At discharge:

$$SOC_{dis} = SOC - Q_m^{-1} \int_{t_i}^{t_i+m} \eta_A(i_a,\tau)^{-1} i_a(t)dt \tag{3}$$

- At charge:

$$SOC_{chg} = SOC + Q_m^{-1} \int_{t_i}^{t_i+m} i_a(t)dt \tag{4}$$

where $SOC_{dis}$ is the electric discharge quantity at discharge mode, $SOC_{chg}$ is the charge quantity of the battery, $Q_m$ is the battery capacity, and $\eta_A(i_a,\tau)$ is the battery's efficiency.

### 2.4 Automated Manual Transmission

The AMT was modeled to change the gear ratio and rotational inertia that correspond to the transmission's gear position. Table 1 shows the gear ratio and reflected rotational inertia that was used in the developed HEV simulator.

| | Gear ratio | Reflected inertia(kg.m$^2$) |
|---|---|---|
| 1st | 3.615 | 0.08999 |
| 2nd | 2.053 | 0.02903 |
| 3rd | 1.393 | 0.00699 |
| 4th | 1.061 | 0.00699 |
| 5th | 0.837 | 0.00699 |

Table 1. Gear ratio of automated manual transmission

The output torque relationships with respect to driving mode are described in Table 2. At Zero Emission Vehicle (ZEV) mode, the electric motor is only actuated when traveling below a critical vehicle speed. In acceleration mode, the power ratio of the motor and the engine is selected in order to meet the demands of the vehicle. At deceleration mode, the regenerative braking torque is produced from the electric motor. The above stated control logic is applied only after considering the SOC of the battery.

| Mode | | Torque relation |
|---|---|---|
| **ZEV** | EV | $T_{out} = T_{motor}$ |
| **Acceleration** | Hybrid | $T_{out} = xT_{motor} + yT_{engine}$ |
| **Deceleration** | Regen. | $T_{out} = T_{regen}$ |
| • Considering the Battery SOC<br>• $x + y = 1$ | | |

Table 2. Output torque relationships with respect to driving mode of AMT-HEV

### 2.5 Vehicle model

When the engine and the electric motor are operating simultaneously, the vehicle state equation is as follows (Yeo et al., 2002)

$$\frac{dV}{dt} = \frac{\dfrac{N_f N_t}{R_t}(T_e + T_m) - F_R}{M + \dfrac{2I_w + (J_e + J_m + J_c)N_t^2 N_f^2 + J_t N_f^2}{R_t^2}} \tag{5}$$

where V is the vehicle velocity, $N_f$ is the final differential gear ratio, $N_t$ is the transmission gear ratio, $R_t$ is the radus of the tire, $F_R$ is the resistance force, $M$ is the vehicle mass, $I_w$ is the equivalent wheel inertia, and $J_e$, $J_m$, $J_c$, and $J_t$ are the inertias of engine, motor, clutch, and transmission, respectively.

## 3. EMB system

The EMB system is environmentally friendly because it does not use a hydraulic system, but rather a 'dry' type Brake-by-wire (BBW) system, which employs an EMB Module (i.e., electric caliper, electro-mechanical disk brake) as the braking module for each wheel. The EMB system is able to provide a large braking force using only a small brake pedal reaction force and a short pedal stroke.

### 3.1 Structure of EMB system

Motors and solenoids can be considered as the electric actuators for EMB systems. The motor is usually chosen as an actuator of the EMB system because the solenoid produces such a small force corresponding to the current input and has such a narrow linear control range that it is unsuitable. In order to generate the proper braking force, Brushless DC

(BLDC) and induction motors are used due to their excellent output efficiency and remarkable durability, respectively. Figure 6 shows a schematic diagram of an EMB system.

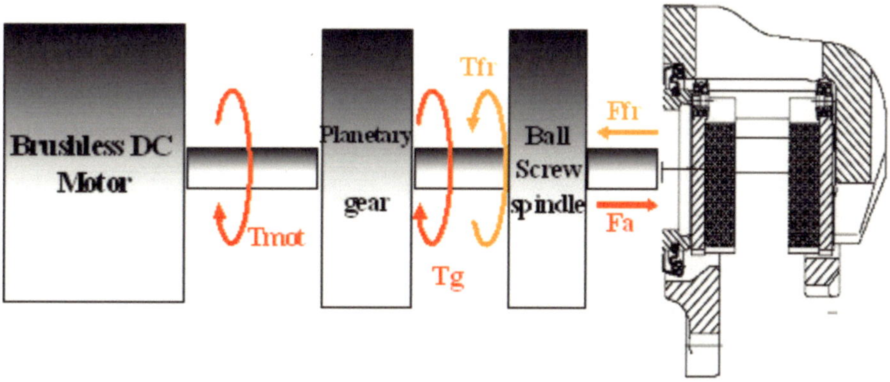


Fig. 6. Schematic diagram of the EMB system

Friction forces are the result of changing resistance of the motor coil and the rigidity of the reduction gear due to temperature fluctuations. To compensate for friction, the control structure for EMB torque adopts a cascade loop. The loop has a low level control logic consisting of the current and velocity control loop shown in Figure 7. This structure requires particularly expensive sensors to measure the clamping force and braking torque; therefore, this paper uses a technique that estimates their values by sensing the voltage, current and position of the DC motor based on the dynamic model of the EMB (Schwarz et al., 1999).

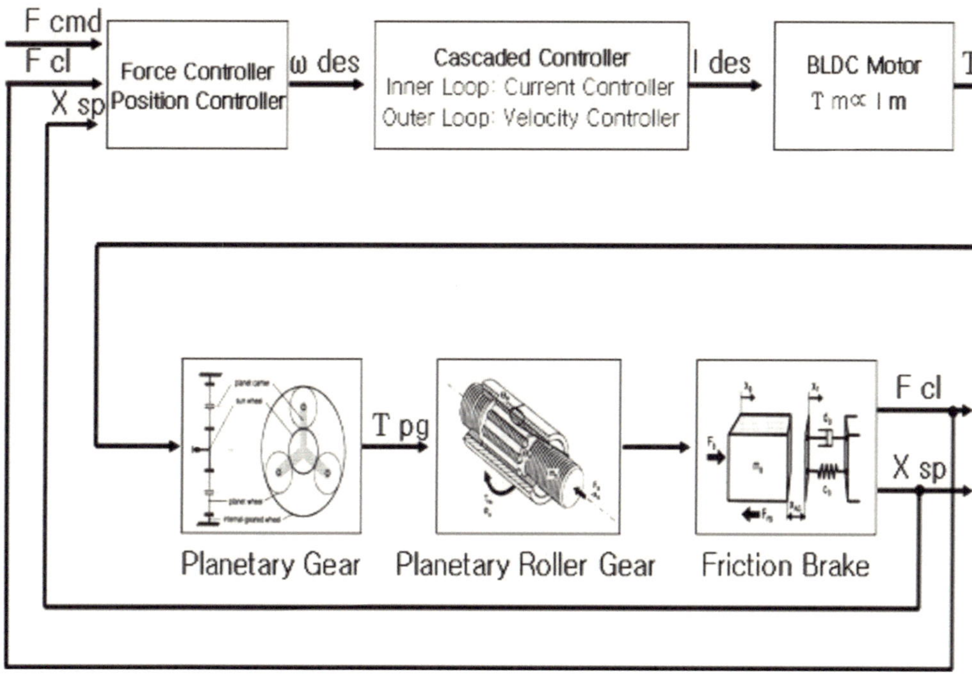


Fig. 7. Control structure of EMB system

### 3.2 Simulation model of EMB system

Figure 8 shows the EMB performance analysis simulator developed in this paper. Force, speed, and electric motor current are fed back via the cascaded loops and controlled by the PID controller.

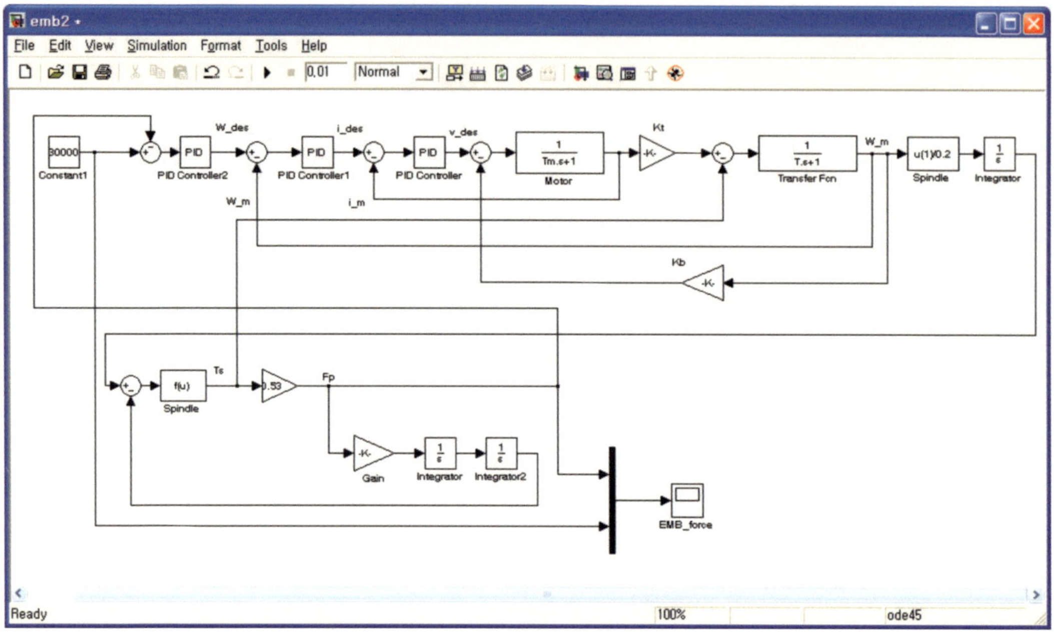


Fig. 8. EMB simulation model

Figure 9 shows the response characteristics of the EMB system. The step response in the time domain is shown at a brake force command of 14 kN.

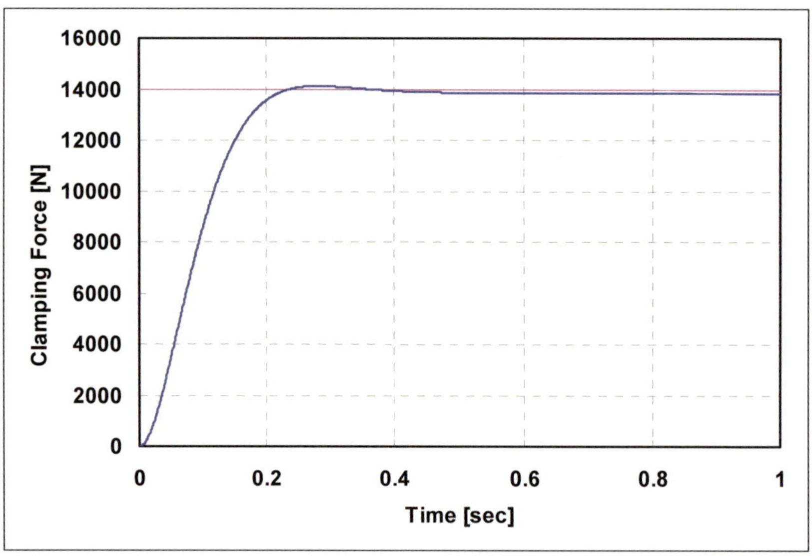


Fig. 9. EMB step response to a force command of 14 kN

## 4. Regenerative braking control algorithm

In conventional vehicles, the energy required to reduce velocity would normally be dissipated and wasted as heat during braking. On the other hand, HEVs have a regenerative braking system that can improve fuel economy. In an HEV, the braking torque is stored in a battery and regenerated through the electric motor/generator (Yaegashi et al., 1998). In this paper, the regenerative braking torque and EMB torque were determined according to the demand of the driver, the characteristics of the electric motor, the SOC of the battery, and the vehicle's velocity. When the regenerative braking power is bigger than the driver's intended braking power, the brake system generates only the regenerative braking torque. When this occurs, the BCU should control the magnitude of regenerative braking torque from the regenerative electric power of motor/generator in order to maintain a brake feeling similar to that of a conventional vehicle (Gao et al., 1999). In this paper, the control algorithm for maximizing regenerative braking torque is performed in order to increase the quantity of battery charge.

### 4.1 Decision logic of regenerative braking torque

Figure 10 shows the flow chart of the control logic for regenerative braking torque.

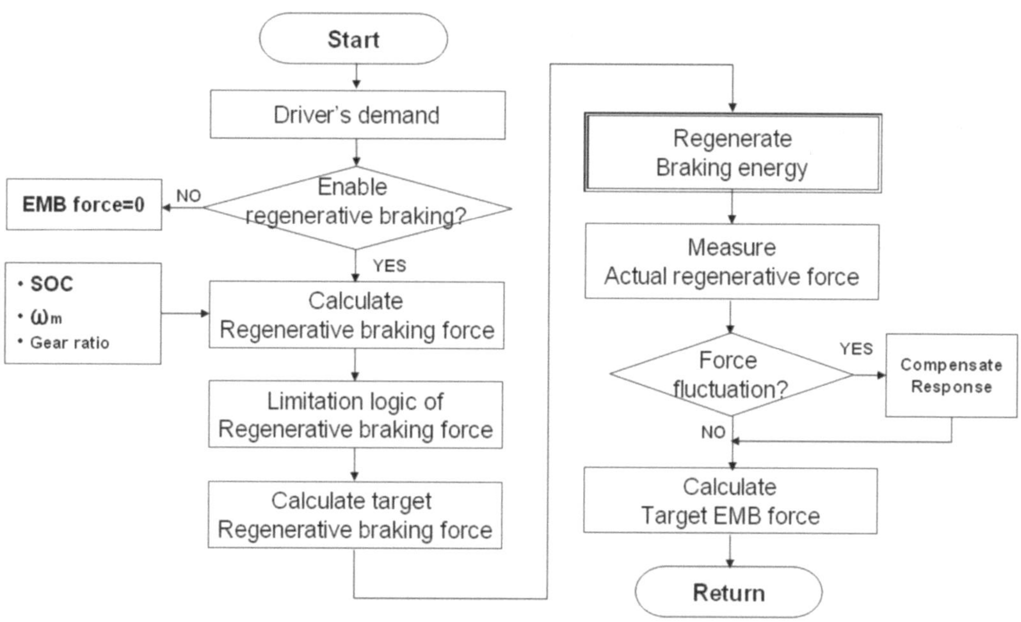


Fig. 10. Regenerative braking control logic flow chart

First, sensing the driver's demand for braking, it calculates the required brake force of the front and rear wheels by using the brake force curve distribution. Then, the logic decides whether the braking system should perform regenerative braking, depending on the states of the accelerator, the brake, the clutch, and the velocity of both engine and vehicle, and on the fail signal. If regenerative braking is available, the optimal force of regenerative braking will subsequently be determined according to the battery's SOC and the speed of the motor. Finally, the algorithm will calculate the target regenerative braking torque. In a situation where the fluctuation of the regenerative braking causes a difference of torque, the response time delay compensation control of the front wheel could be used to minimize the fluctuation of the target brake force. After the target braking torque is determined, the remainder of the difference between target braking torque and the regenerative braking torque will be transmitted via the EMB system.

### 4.2 Limitation logic of regenerative braking torque

Overcharging the battery during regenerative braking reduces battery durability. Therefore, when the SOC of the battery is in the range of 50%-70%, the logic applies the greatest regenerative torque; however, when the SOC is above 80%, it does not perform regeneration (Yeo et al., 2004).

## 5. HEV performance simulator using MATLAB/Simulink

The brake performance simulator was created for validating the regenerative braking control logic of the parallel HEV. The modeling of the HEV powertrain (including the engine, the motor, the battery, the automated manual transmission, and EMB) was performed, and the control algorithm for regenerative braking was developed using MATLAB/Simulink. Figure 11 illustrates the AMT-HEV simulator.

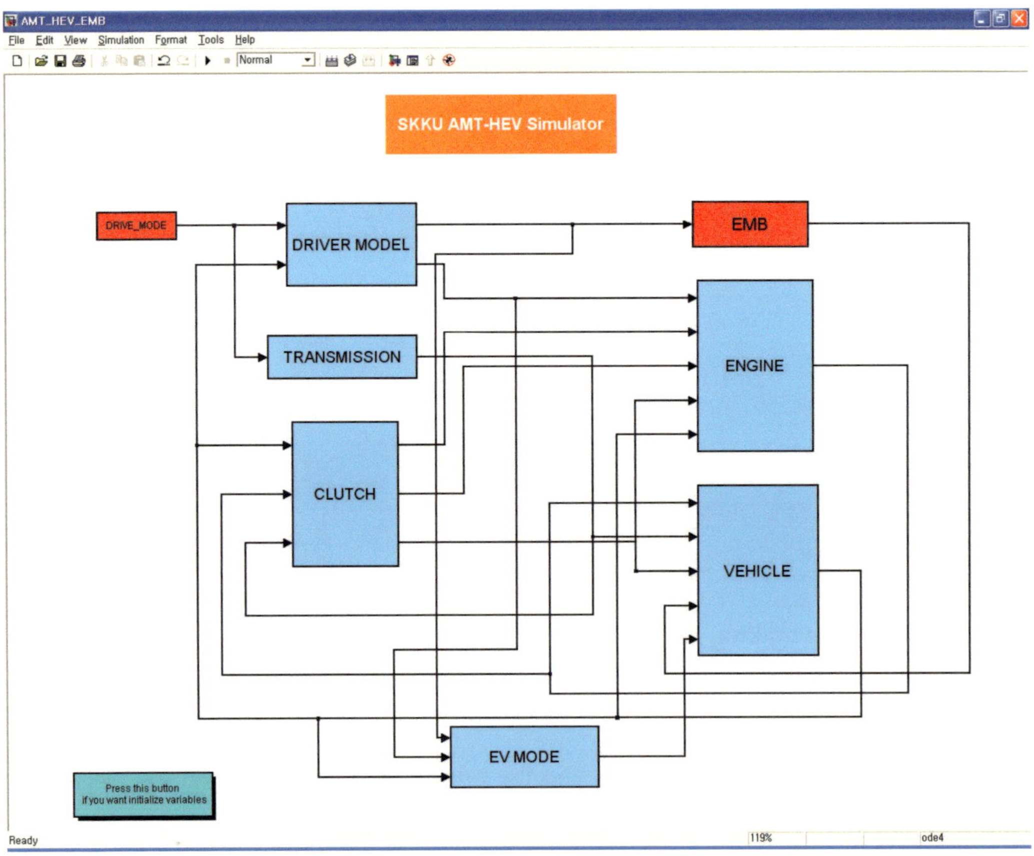


Fig. 11. AMT-HEV simulator with EMB

## 6. Simulation results

The simulation results for the Federal Urban Drive Schedule (FUDS) mode using the performance simulator are shown in Figure 12.
According to Figure 12, the brake pedal and accelerator positions are changing relative to the drive mode. Subsequently, the vehicle's velocity successfully chases the drive mode. The torque of the engine and the motor is illustrated in the figure. The graph of battery SOC adequately shows charging state by regenerative braking during deceleration.

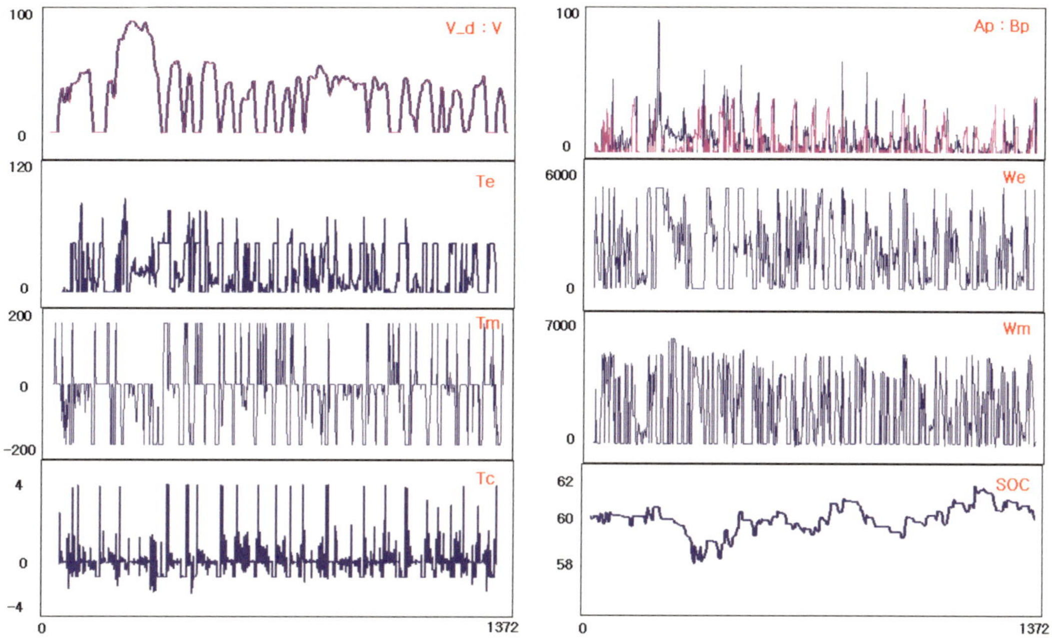


Fig. 12. Simulation results for FUDS mode

## 7. Conclusion

In this paper, the performance simulation for a hybrid electric vehicle equipped with an EMB system was conducted. A performance simulator and dynamics models were developed to include such subsystems as the engine, the motor, the battery, AMT, and EMB. The EMB control algorithm that applied the PID control technique was constructed based on cascade control loops composed of the current, velocity, and force control systems. The simulation results for FUDS mode showed that the HEV equipped with an EMB system can regenerate the braking energy by using the proposed regenerative braking control algorithm.

## 8. References

Ahn, J., Jung, K., Kim, D., Jin, H., Kim, H. and Hwang, S. (2009). Analysis of a regenerative braking system for hybrid electric vehicles using an electro-mechanical brake, *Int. J. of Automotive Technology*, Vol. 10(No. 2): 229–234.

Emereole, O. and Good, M. (2005). The effect of tyre dynamics on wheel slip control using electromechanical brakes. *SAE Paper No.* 2005-01-0419.

Gao, Y., Chen, L. and Ehsani, M. (1999). Investigation of the effectiveness of regenerative braking for EV and HEV. *SAE Paper No.* 1999-01-2910.

Kim, D., Hwang, S. and Kim, H. (2008). Vehicle stability enhancement of four-wheel-drive hybrid electric vehicle using rear motor control, *IEEE Transactions on Vehicular Technology*, Vol. 57(No. 2): 727-735.

Line, C., Manzie, C. and Good, M. (2004). Control of an electromechanical brake for automotive brake-by-wire systems with an adapted motion control architecture. *SAE Paper No.* 2004-01-2050.

Nakamura, E., Soga, M., Sakaki, A., Otomo, A. and Kobayashi, T. (2002). Development of electronically controlled brake system for hybrid vehicle. *SAE Paper No.* 2002-01-0900.

Peng, D., Zhang, Y., Yin, C.-L., and Zhang, J.-W. (2008). Combined control of a regenerative braking and antilock braking system for hybrid electric vehicles, *Int. J. of Automotive Technology*, Vol. 9(No. 6): 749-757.

Schwarz, R., Isermann, R., Bohm, J., Nell, J. and Rieth, P. (1999). Clamping force estimation for a brake-by-wire actuator. *SAE Paper No.* 1999-01-0482.

Semm, S., Rieth, P., Isermann, R. and Schwarz, R. (2003). Wheel slip control for antilock braking systems using brake-by-wire actuators. *SAE Paper No.* 2003-01-0325.

Yaegashi, T., Sasaki, S. and Abe, T. (1998). Toyota hybrid system: It's concept and technologies. *FISITA* F98TP095.

Yeo, H. and Kim, H. (2002). Hardware-in-the-loop simulation of regenerative braking a hybrid electric vehicle. *Proc. Instn. Mech. Engrs.*, Vol. 216: 855-864.

Yeo, H., Song, C., Kim, C. and Kim, H. (2004). Hardware in the loop simulation of hybrid electric vehicle for optimal engine operation by CVT ratio control. *Int. J. of Automotive Technology*, Vol. 5(No. 3): 201-208.

# 9

# Control of Electric Vehicle

Qi Huang, Jian Li and Yong Chen
*University of Electronic Science and Technology of China*
*P.R.China*

## 1. Introduction

The major components of an electric vehicle system are the motor, controller, power supply, charger and drive train (wry, 2003). Fig. 1 demonstrates a system model for an electric vehicle. Controller is the heart of an electric vehicle, and it is the key for the realization of a high-performance electric vehicle with an optimal balance of maximum speed, acceleration performance, and traveling range per charge.

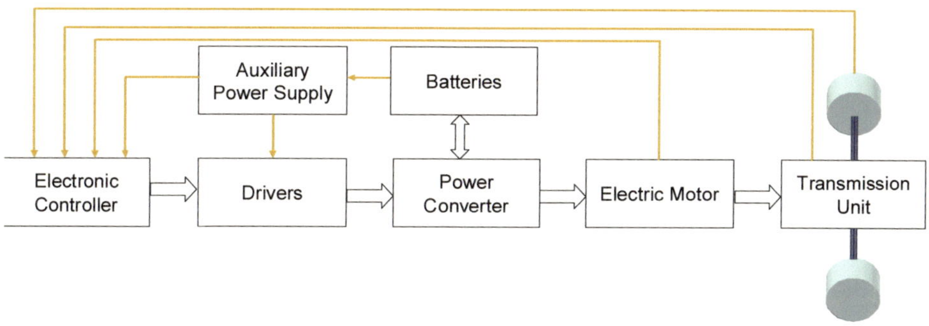


Fig. 1. Major components in an electric vehicle

Control of Electric Vehicle (EV) is not a simple task in that operation of an EV is essentially time-variant (e.g., the operation parameters of EV and the road condition are always varying). Therefore, the controller should be designed to make the system robust and adaptive, improving the system on both dynamic and steady state performances. Another factor making the control of EV unique is that EV's are really "energy-management" machines (Cheng et al., 2006). Currently, the major limiting factor for wide-spread use of EV's is the short running distance per battery charge. Hence, beside controling the performance of vehicle (e.g., smooth driving for comfortable riding), significant efforts have to be paid to the energy management of the batteries on the vehicle.
However, from the viewpoint of electric and control engineering, EV's are advantagous over traditional vehicles with internal combustion engine. The remarkable merit of EV's is the electric motor's excellent performance in motion control, which can be summerized as (Sakai & Hori, 2000): (1) torque generation is very quick and accurate, hence electric motors can be controlled much more quickly and precisely; (2) output torque is easily comprehensible; (3) motor can be small enough to be attached to each wheel; (4) and the controller can be easily designed and implemented with comparatively low cost.

Hence, in recent years, there is quite a lot of researches in the exploring advanced controll strategies in electric vehicles. As the development of the high computing capability microprocessor, such as DSP (Digital Signal Processor), it is possible to perform complex control on the electric vehicle to achieve optimal performance (Liu et al., 2004). These capabilities can be utilized to enhance the performance and safety of individual vehicles as well as to operate vehicles in formations for specific purposes (Lin & Kanellakopoulos, 1995). Due to the complex operation condition of electric vehicle, intelligent or fuzzy control is generally used to increase efficiency and deal with complex operation modes (Poorani et al., 2003; Khatun et al., 2003). However, it is essential to establish a model-based control for the EV system, and systematically study the characteristics to achieve optimal and robust control. This chapter will mainly focuses on model-based control design for EV's and the implementation of the platform for realization of variant control strategies.

## 2. Modeling of electric vehicle

Generally, the modelling of an EV involves the balance among the forces acting on a running vehicle, as shown in Fig. 2. The forces are categorized into road load and tractive force. The road load consists of the gravitational force, hill-climbing force, rolling resistance of the tires and the aerodynamic drag force. Consider all these factors, a vehicle dynamic model that governs the kinetics of the wheels and vehicle can be written as (wry, 2003):

$$F = \mu_{rr} mg + \frac{1}{2}\rho A C_d v^2 + mg\sin\phi + m\frac{dv}{dt} \tag{1}$$

Where, $m$ is the mass of the electric vehicle; $g$ is the gravity acceleration; $v$ is the driving velocity of the vehicle; $\mu_{rr}$ is the rolling resistance coefficient; $\rho$ is the air density; $A$ is the frontal area of the vehicle; $C_d$ is the drag coefficient; and $\phi$ is the hill climbing angle.

The rolling resistance is produced by the flattening of the tire at the contact surface of the roadway. The main factors affecting the rolling resistance coefficient $\mu_{rr}$ are the type of tyre and the tyre pressure. It is generally obtained by measurement in field test. The typical range is 0.005-0.015, depending on the type of tyre. The rolling resistance can be minimized by keeping the tires as much inflated as possible.

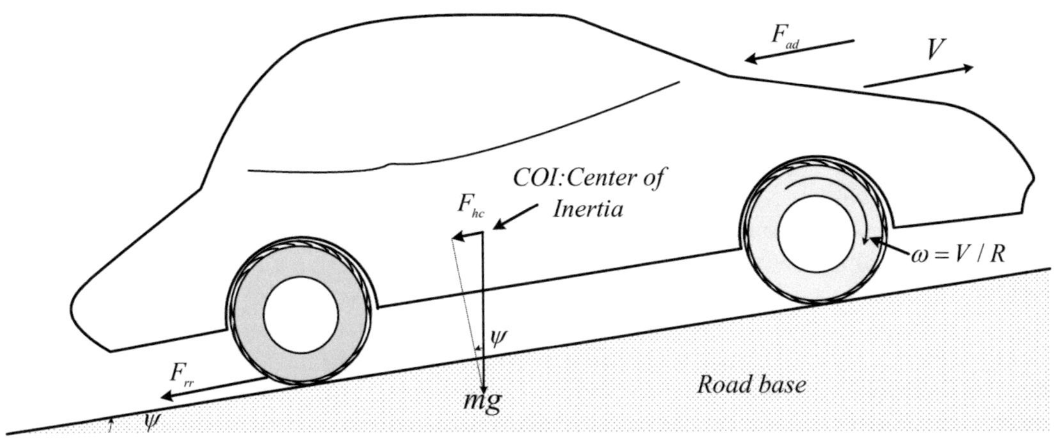


Fig. 2. External forces applied on a running vehicle

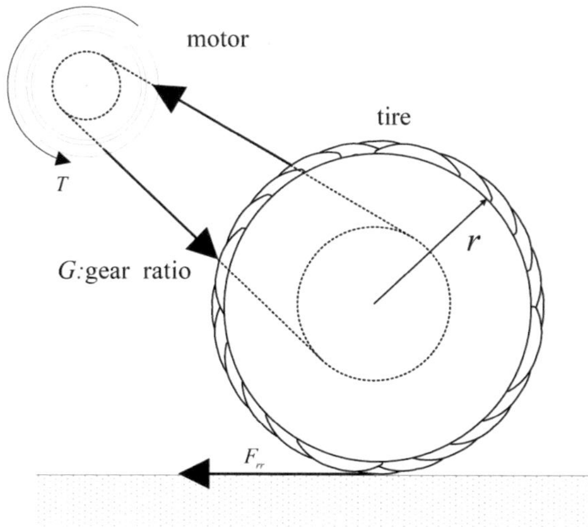


Fig. 3. A simplified model for motor driving tyre

In equation (1), the first term corresponds to the rolling resistance force; the second term corresponds to the aerodynamic drag force; the third term corresponds to the hill climbing force; and the forth term corresponds to the acceleration force.

This resultant force *F*, will produce a counteractive torque to the driving motor, i.e., the tractive force. For vibration study, the connection between the driving motor and the tyre should be modeled in detail. Interested readers are refered to (Profumo et al., 1996). In this chapter, a simplified model, as shown in Fig. 3, will be used. With this simplified model, the relationship between the tractive force and the torque produced by the motor can be obtained as:

$$T_L = F \cdot \frac{r}{G} \tag{2}$$

Where $r$ is the tyre radius of the electric vehicle, $G$ is the gearing ratio, and $T_L$ is the torque produced by the driving motor.

## 3. Electric motor and their models

Presently, brushed DC motor, brushless DC motor, AC induction motor, permanent magnet synchronous motor (PMSM) and switched reluctance motor (SRM) are the main types of motors used for electric vehicle driving (Chan, 1999). The selection of motor for a specific electric vehicle is dependent on many factors, such as the intention of the EV, ease of control, etc.

In control of electric vehicle, the control objective is the torque of the driving machine. The throttle position and the break is the input to the control system. The control system is required to be fast reponsive and low-ripple. EV requires that the driving electric machine has a wide range of speed regulation. In order to guarantee the speed-up time, the electric machine is required to have large torque output under low speed and high over-load capability. And in order to operate at high speed, the driving motor is required to have certain power output at high-speed operation. In this chapter, the former four types of motors that can be found in many applications will be discussed in detail.

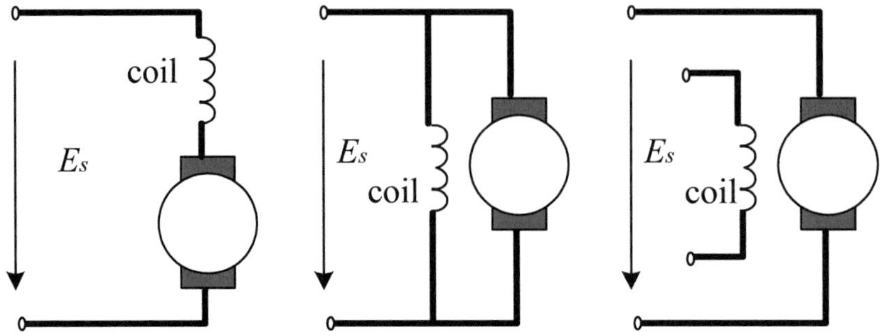


Fig. 4. Three types of winding configuration in DC motor

### 3.1 Brushed DC motor

Due to the simplicity of DC motor controlling and the fact that the power supply from the battery is DC power in nature, DC motor is popularly selected for the traction of electric vehicles. There are three classical types of brushed DC motor with field windings, series, shunt and 'separately excited' windings, as shown in Fig. 4. The shunt wound motor is particularly difficult to control, as reducing the supply voltage also results in a weakened magnetic field, thus reducing the back EMF, and tending to increase the speed. A reduction in supply voltage may, in some circumstances, have very little effect on the speed. The separately excited motor allows one to have independent control of both the magnetic flux and the supply voltage, which allows the required torque at any required angular speed to be set with great flexibility. A series wound DC motor is easy to use and with added benefit of providing comparatively larger startup torque. A (series) DC motor can generally be modeled as (Mehta & Chiasson, 1998):

$$\begin{cases} (L_a + L_{field})\dfrac{di}{dt} = V - (R_a + R_f)i - L_{af}i \cdot \omega \\ J\dfrac{d\omega}{dt} = L_{af}i^2 - B\omega - T_L \end{cases} \tag{3}$$

Where: $i$ is the armature current (also field current); $\omega$ is the motor angular speed; $L_a$, $R_a$, $L_{field}$, $R_f$ are the armature inductance, armature resistance, field winding inductance and field winding resistance respectively; $V$ is the input voltage, as the control input; $L_{af}$ is the mutual inductance between the armature winding and the field winding, generally non-linear due to saturation; $J$ is the inertia of the motor, including the gearing system and the tyres; $B$ is the viscous coefficient; and $T_L$ is representing the external torque, which is quantitatively the same as the one aforementioned.

### 3.2 Brushless DC motor

The disadvantages of brushed DC motor are its frequent maintenance and low life-span for high intensity uses. Therefore, brushless DC (BLDC) motor is developed. Brushless DC motors use a rotating permanent magnet in the rotor, and stationary electrical magnets on the motor housing. A motor controller converts DC to AC. This design is simpler than that of brushed motors because it eliminates the complication of transferring power from outside the motor to the spinning rotor. Brushless motors are advantageous over brushed ones due to their long life span, little or no maintenance, high efficiency, and good performance of timing. The disadvantages are high initial cost, and more complicated motor speed controllers (Wu et al., 2005).

A BLDC motor is composed of the motor, controller and position sensor. In the BLDC motor, the electromagnets do not move; instead, the permanent magnets rotate and the armature remains static. The rotor magnetic steel is radially placed, and the permanent magnets (generally Neodymium-iron-boron: NdFeB) are installed on the surface. The magnetic permeability of such permanent magnets is close to that of air, hence can be regarded as part of the air gap. Hence, there is no salient pole effect, so that the magnetic field across the air gap is uniformly distributed. The position sensor functions like the commutator of brushed DC motor, reflecting the position of the rotor and determining the phase of current and space distribution of magnetic force.

The BLDC motor is actually an AC motor. The wires from the windings are electrically connected to each other either in delta configuration or wye ("Y"-shaped) configuration. In Fig. 5, an equivalent circuit of wye-connected BLDC is shown. With this configuration, the simplified model can be obtained as:

$$\begin{bmatrix} u_a \\ u_b \\ u_c \end{bmatrix} = \begin{bmatrix} R & 0 & 0 \\ 0 & R & 0 \\ 0 & 0 & R \end{bmatrix} \begin{bmatrix} i_a \\ i_b \\ i_c \end{bmatrix} + \begin{bmatrix} L & 0 & 0 \\ 0 & L & 0 \\ 0 & 0 & L \end{bmatrix} \cdot P \begin{bmatrix} i_a \\ i_b \\ i_c \end{bmatrix} + \begin{bmatrix} e_a \\ e_b \\ e_c \end{bmatrix} + \begin{bmatrix} U_N \\ U_N \\ U_N \end{bmatrix} \tag{4}$$

Where, $L = L_S - M$; $L_S$: self-inductance of the windings; $M$: mutual inductance between two windings; $R$: stator resistance per phase; $u_a, u_b, u_c$ : stator phase voltages; $i_a, i_b, i_c$ : stator phase currents; $e_a, e_b, e_c$ : the back emfs in each phase; $P = \frac{d}{dt}$.

The generated electromagnetic torque is given by:

$$T_e = \frac{P_e}{\omega} = (e_a i_a + e_b i_b + e_c i_c) / \omega \tag{5}$$

And the kinetics of the motor can be described as:

$$T_e - T_L - f\omega = J\frac{d\omega}{dt} \tag{6}$$

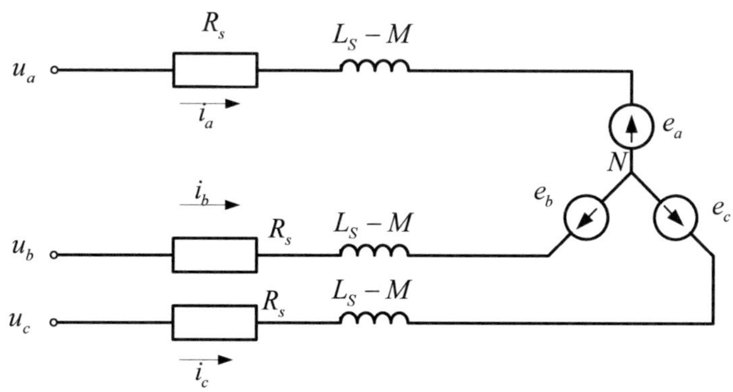

Fig. 5. Equivalent circuit of BLDC

Where, $\omega$: the angular velocity of the motor; $T_e$, $T_L$: electromagnetic torque of the motor and the load torque; $P_e$: electromagnetic power of the motor; $J$: moment of inertia; $f$: friction coefficient.
Under normal operation, only two phases are in conduction. Then the voltage balance equation, back EMF equation, torque equation, and kinetic equations that govern the operation of a WYE connected BLDC motor can be obtained as:

$$\begin{cases} u_d = E + i \cdot R + L \cdot \dfrac{di}{dt} \\ E = K_e \cdot n \\ T_e = K_T \cdot n \\ T_e = T_L + f\omega + J\dfrac{d\omega}{dt} \end{cases} \tag{7}$$

Where, $u_d$: the voltage across the two windings under conduction; $E$: the back EMF of the two windings under conduction; $K_T$: torque coefficient, and $K_e$: back EMF coefficient.
It is shown that, in BLDC motors, current to torque and voltage to rpm are linear relationships.

### 3.3 Permanent magnet synchronous motor

A permanent magnet synchronous motor is a motor that uses permanent magnets to produce the air gap magnetic field rather than using electromagnets. Such motors have significant advantages, such as high efficiency, small volume, light weight, high reliability and maintenance-free, etc., attracting the interest of EV industry. The PMSM has a sinusoidal back emf and requires sinusoidal stator currents to produce constant torque while the BDCM has a trapezoidal back emf and requires rectangular stator currents to produce constant torque. The PMSM is very similar to the wound rotor synchronous machine except that the PMSM that is used for servo applications tends not to have any damper windings and excitation is provided by a permanent magnet instead of a field winding. Hence the d, q model of the PMSM can be derived from the well known model of the synchronous machine with the equations of the damper windings and field current dynamics removed.
In a PMSM, the magnets are mounted on the surface of the motor core. They have the same role as the field winding in a synchronous machine except their magnetic field is constant and there is no control on it. The stator carries a three-phase winding, which produces a near sinusoidal distribution of magneto motive force based on the value of the stator current. In modeling of rotating machines like PMSM, it is a general practice to perform Park transform and deal the quantities under dq framework. The dqo transform applied to a three-phase quantities has following form:

$$\begin{bmatrix} x_d \\ x_q \\ x_0 \end{bmatrix} = \sqrt{\frac{2}{3}} \cdot \begin{bmatrix} \cos\theta & \cos(\theta - \frac{2\pi}{3}) & \cos(\theta + \frac{2\pi}{3}) \\ -\sin\theta & -\sin(\theta - \frac{2\pi}{3}) & -\sin(\theta + \frac{2\pi}{3}) \\ \sqrt{\frac{1}{2}} & \sqrt{\frac{1}{2}} & \sqrt{\frac{1}{2}} \end{bmatrix} \cdot \begin{bmatrix} x_a \\ x_b \\ x_c \end{bmatrix} \tag{8}$$

Where, the $x$ can be voltage $u$ or current $i$.
Under the dq0 framework, the equivalent circuit of d-axis and q-axis circuits of a PMSM motor is shown in Fig. 6, and the model of a PMSM can be written as (Cui et al., 2001):

$$\begin{cases} u_d = r_s i_d + p\Psi_d - \omega\Psi_q \\ u_q = r_s i_q + p\Psi_q - \omega\Psi_d \\ \Psi_d = L_d i_d + \Psi \\ \Psi_q = L_q * i_q \\ T_e = 1.5 p_m \left[ \Psi i_q + (L_d - L_q) i_d i_q \right] \\ J\dfrac{d\omega_r}{dt} = T_e - B\omega_r - T_L \end{cases} \tag{9}$$

Where, $\Psi_d$, $\Psi_q$: the flux linkages of d-axis and q-axis respectively; $L_d$, $L_q$: self inductance of dq axes; $i_d$, $i_q$: dq-axis current; $u_d$, $u_q$: dq-axis voltage; $\omega_r$: angular velocity of rotor; $r_s$: stator resistance; $p_m$: number of poles; $\Psi$: flux linkage produced by the rotor permanent magnet; $T_e$: motor torque; $T_L$: load torque; $J$: moment of inertia; $B$: friction coefficient; $p$: differential operator.
The first two equations are the equations for stator voltage, the next two equations are about the magnetic flux linkage, the fifth equation is about the calculation of torque, and the last equation is about the kinetics of the motor.

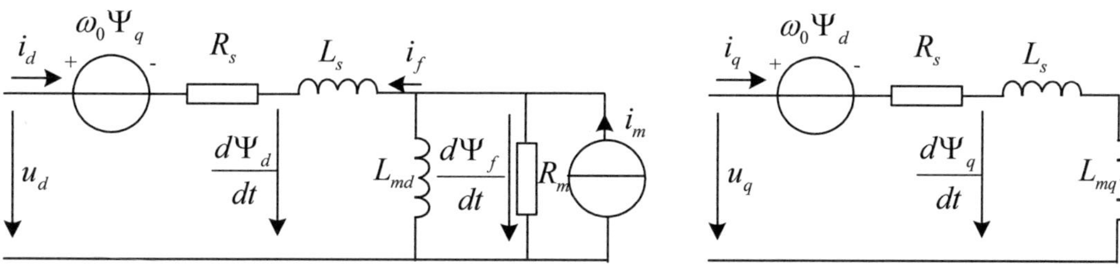


Fig. 6. d-axis and q-axis equivalent circuit model of PMSM

### 3.4 Induction motor

Induction machines are among the top candidates for driving electric vehicles and they are widely used in modern electric vehicles. Some research even concludes that the induction machine provides better overall performance compared to the other machines (Gosden et al., 1994).
An induction motor (or asynchronous motor or squirrel-cage motor) is a type of alternating current motor where power is supplied to the rotor by means of electromagnetic induction. It has the advantages such as low-cost, high-efficiency, high reliability, maintenance-free, easy for cooling and firm structure, etc. making it specially competitive in EV driving. In induction motor, stator windings are arranged around the rotor so that when energised with a polyphase supply they create a rotating magnetic field pattern which sweeps past the rotor. This changing magnetic field pattern induces current in the rotor conductors, which interact with the rotating magnetic field created by the stator and in effect causes a rotational motion on the rotor.

AC induction motor is a time-varying multi-variable nonlinear system, hence the modeling task is not easy. For simplicity, following assumptings have to be made:

- Magnetic circuit is linear, and saturation effect is neglected;
- Symmetrical two-pole and three phases windings (120° difference) with edge effect neglected;
- Slotting effects are neglected, and the flux density is radial in the air gap and distributed along the circumference sinusoidally;
- Iron losses are neglected.

With such assumptions, the physical model of an induction motor can be given as shown in Fig. 7. The three-phase stators are fixed on A, B and C axes, which are stationary reference frames. The three-phase rotor windings are fixed on a, b and c axes, which are rotating frames. Hence the equations governing the dynamics of the induction motor can be given as (Dilmi & Yurkovich, 2005):

$$u = Ri + L\frac{di}{dt} + \omega_r \frac{\partial L}{\partial \theta_0} i \tag{10}$$

$$\frac{\partial^2 \theta_0}{\partial t^2} = \frac{\partial \omega_r}{\partial t} = \frac{1}{J}(T - T_L) = \frac{1}{J}(\frac{1}{2} i^T \frac{\partial L}{\partial \theta_0} i - T_L) \tag{11}$$

Where, $u = [u_A, u_B, u_C, u_a, u_b, u_c]^T$, vector of stator and rotor voltages; $i = [i_A, i_B, i_C, i_a, i_b, i_c]^T$, vector of stator and rotor current; $\omega_r = d\theta_0 / dt$: the angular speed of rotation; $J$: the total moment of inertia; $T_L$: load torque; $R = diag[R_1 \quad R_1 \quad R_1 \quad R_2 \quad R_2 \quad R_2]$: where $R_1$ is the resistance of stator winding and $R_2$ is the resistance of rotor winding;

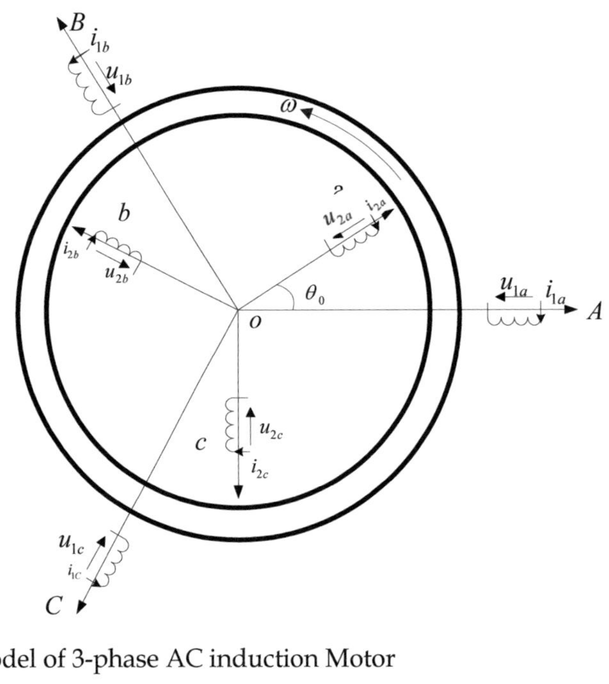


Fig. 7. Physical Model of 3-phase AC induction Motor

$L = \begin{bmatrix} L_{11} & L_{12} \\ L_{21} & L_{22} \end{bmatrix}$, where $L_{11} = \begin{bmatrix} L_A & L_{AB} & L_{AB} \\ L_{AB} & L_A & L_{AB} \\ L_{AB} & L_{AB} & L_A \end{bmatrix}$, $L_{22} = \begin{bmatrix} L_a & L_{ab} & L_{ab} \\ L_{ab} & L_a & L_{ab} \\ L_{ab} & L_{ab} & L_a \end{bmatrix}$ and $L_{12} = L_{12}^T = M \begin{bmatrix} \cos\theta & \cos(\theta - 120^\circ) & \cos(\theta + 120^\circ) \\ \cos(\theta - 120^\circ) & \cos\theta & \cos(\theta - 120^\circ) \\ \cos(\theta + 120^\circ) & \cos(\theta - 120^\circ) & \cos\theta \end{bmatrix}$; $L_A$, $L_a$: self-inductance of stator and rotor; $L_{AB}$, $L_{ab}$: mutual inductance of stators and rotors respectively; $M$: mutual inductance between stator and rotor.

The first equation is the voltage equation and the second equation is the kinetic equation of the motor.

## 4. Controller design of electric vehicle driven by different motors

Fig. 8 shows a universal framework for electric vehicle controller. The vehicle is driven by a motor, which is supplied by the battery through a controlled power circuit. Other than circuit for control of the motor, there is quite a lot of auxiliary control for auto electronics. The control strategies are implemented in the microprocessor, such as DSP (Digital Signal Processor).

Control of electric vehicle is essentially the control of motor. In Fig. 8, only the motor and its associated driving power circuit will be replaced with different motors. With different motors, it is necessary to use different control strategies. However, it is not possible to include all type of motor and control strategies in one book. Hence, in this chapter, only one typical controller or control strategy will be presented. It is noticed that (Chan, 1999) generally PWM control is used for DC motor, while variable-voltage variable-frequency (VVVF), FOC (field-oriented control) and DTC (direct torque control) are used for induction motor. And some traditional control algorithms, such as PID, cannot satisfy the requirements of EV control. Many modern high-performance control technologies, such as adaptive control, fuzzy control, artificial neuro network and expert system are being used in EV controllers.

### 4.1 Driven by brushed DC motor

In this subsection, the controller design for an EV driven by series wound DC motor will be discussed. When the electric vehicle is driven by a series wound DC motor, the overall system model is the combination of (1) and (3):

$$\begin{cases} (L_a + L_{field})\dfrac{di}{dt} = V - (R_a + R_f)i - L_{af}i \cdot \omega \\ (J + m\dfrac{r^2}{G^2})\dfrac{d\omega}{dt} = L_{af}i^2 - B\omega - \dfrac{r}{G}(\mu_{rr}mg + \dfrac{1}{2}\rho AC_d v^2 + mg\sin\phi) \end{cases} \tag{12}$$

In this case, a model-based controller can be designed. Unlike other applications in which the system generally operates around the equilibrium point, the operation of EV may take a very wide range (e.g., from zero to full speed). Hence, it is essentially to design EV controller with nonlinear control techniques. The model-based controller is very sensitive to the uncertainties in the parameters. Many parameters in the complex vehicle dynamics

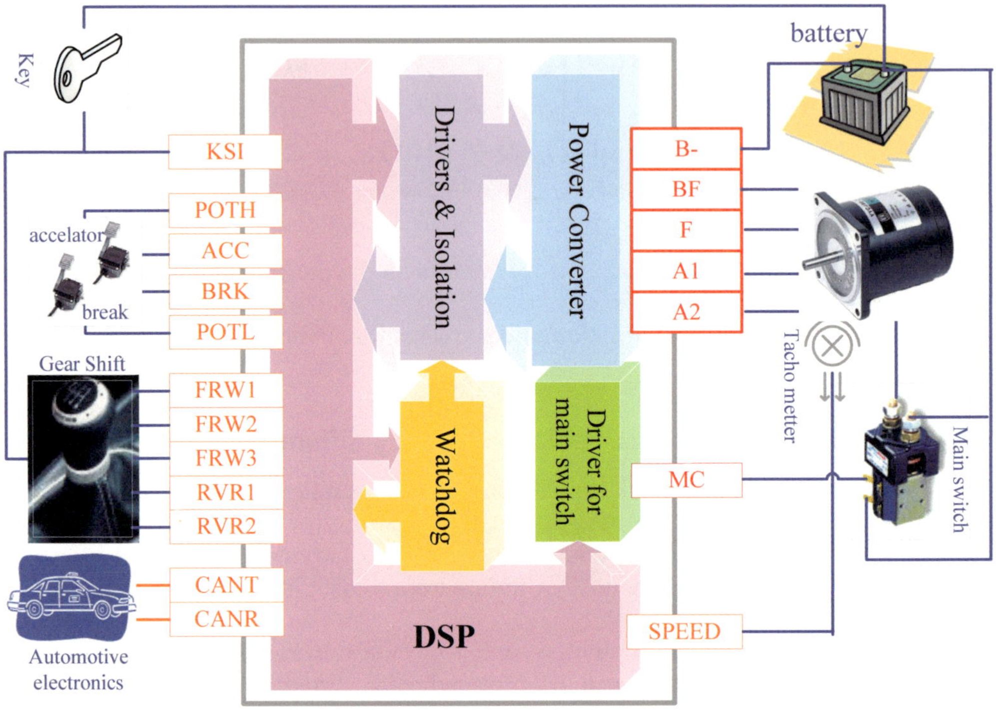


Fig. 8. Model of electric vehicle controller

cannot be precisely modeled and some parameters may vary due to the varying operation conditions. For example, the resistance in the armature winding of a motor would change as the operation temperature varies. Hence, when designing the controller, the robustness of the controller should be first considered. In this subsection, a nonlinear robust and optimal controller (Huang et al., 2009) will be discussed.

For the convenience of designing nonlinear controller, first change the model in (12) into the following format:

$$\begin{cases} \dot{X} = f(X) + g(X)u \\ y = h(X) \end{cases} \tag{13}$$

Where:

$$X = \begin{bmatrix} x_1 \\ x_2 \end{bmatrix} = \begin{bmatrix} i \\ \omega \end{bmatrix}; f(X) = \begin{bmatrix} -\dfrac{R_a + R_f}{L_a + L_{field}} x_1 - \dfrac{L_{af}}{L_a + L_{field}} x_1 \cdot x_2 \\ \dfrac{1}{J + m\dfrac{r^2}{G^2}} \left\{ L_{af} x_1{}^2 - B x_2 - \dfrac{r}{G}(\mu_{rr} mg + \dfrac{1}{2}\rho A C_d \dfrac{r^2}{G^2} x_2{}^2 + mg\sin\varphi) \right\} \end{bmatrix};$$

$$g(X) = \begin{bmatrix} \dfrac{1}{L_a + L_{field}} \\ 0 \end{bmatrix}; \; h(X) = x_2 .$$

In order to consider the uncertainties of the system, further change the form of (13) into:

$$\begin{cases} \dot{X} = f(X) + q(X,\theta(t)) + g(X)u \\ y = h(X) \end{cases} \tag{14}$$

$f(X)$, $g(X)$ and $h(X)$ are the same as in previous section. $q(X,\theta(t))$ is used to include the model uncertainties, where $\theta(t)$ is the uncertainty vector. In the model described above, the rolling resistance coefficient and aerodynamic dragging coefficients cannot be precisely modeled. These coefficients are always varying along the moving of the vehicle (e.g., due to wind). Also, the resistance of the windings is also varying due to the variation of temperature. Hence, $q(X,\theta(t))$ can be modeled as:

$$q(X,\theta(t)) = \begin{bmatrix} \dfrac{i}{L_a + L_{field}} \Delta R \\ -\dfrac{r/G}{J + m\dfrac{r^2}{G^2}} (mg\Delta\mu_{rr} + \dfrac{r^2}{G^2}\omega^2 \Delta C_{ad}) \end{bmatrix} \tag{15}$$

Where, $\Delta R$, $\Delta\mu_{rr}$ and $\Delta C_{ad}$ are the uncertainties in winding resistance, rolling resistance coefficient and aerodynamic dragging coefficient respectively, with $\Delta R_m$, $\Delta\mu_{rr\_m}$ and $\Delta C_{ad\_m}$ representing their maximum uncertainties.
By using the similar coordinate transformation, we have,

$$\begin{cases} \dot{z}_1 = \dot{\omega} - \dot{\omega}_0 = z_2 + L_q h(X) \\ \dot{z}_2 = L_f^2 h(X) + L_g L_f h(X) u + L_q L_f h(X) \end{cases} \tag{16}$$

Where,

$$L_q h(X) = -\frac{r/G}{J + m\frac{r^2}{G^2}} (mg\Delta\mu_{rr} + \frac{r^2}{G^2} x_2^2 \Delta C_{ad})$$

$$L_q L_f h(X) = \frac{1}{J + m\frac{r^2}{G^2}} \frac{2x_1^2}{L_a + L_{field}} \Delta R - \frac{r/G(mg\Delta\mu_{rr} + \frac{r^2}{G^2} x_2{}^2 \Delta C_{ad})(\rho A C_d \frac{r^2}{G^2} x_2 - B)}{\left(J + m\frac{r^2}{G^2}\right)^2}$$

Hence, by using the recursive backstepping design method with robust control system (Marino & Tomei, 1993; Freeman & Kokotovic, 1996), one can select a robust control Lyapunov function (rclf) as:

$$V(z_1, z_2) = \frac{1}{2} z_1^2 + \frac{1}{2}(z_2 - \tilde{z}_2)^2 \tag{17}$$

Where, $\tilde{z}_2 = s_1 z_1$, and $s_1(z_1) = \dfrac{r^3/G^3}{J + m\dfrac{r^2}{G^2}} (z_1 + 2\omega_0) \Delta C_{ad\_m} - 2$.

Then, the control law in (18) can robustly stabilize the electric vehicle system with any parametric uncertainties:

$$u = \frac{\tilde{z}_3 - L_f^2 h(X)}{L_g L_f h(X)} \tag{18}$$

Where,

$$\tilde{z}_3 = s_2(z_2 - s_2 z_1) \tag{19}$$

$$\begin{aligned} s_2 = &\frac{T_1^2}{2} - 1 - T_1 - \beta_6 \Delta R_m - \{T_1[3 + 2\alpha_2 z_1 + 4\alpha_2\omega_0]\Delta C_{ad\_m} \\ &- \beta_6 \Delta R_m s_1 - 1 + \left[\left(\alpha_4 z_1^2 + (\alpha_3 + 3\alpha_4\omega_0)z_1 + (2\alpha_3\omega_0 + 3\alpha_4\omega_0^2)\right)\Delta C_{ad\_m} \right. \\ &\left. + \alpha_2 \Delta\mu_{rr\_m} + (\alpha_7 z_1 + \alpha_8)\Delta R_m \ \right] \}^2 / 4 \end{aligned} \tag{20}$$

$T_1 = (2\alpha_1 z_1 + 2\alpha_1\omega_0)\Delta C_{ad\_m} + 2$; $\alpha_1 = \left(r^3 / G^3\right) / \left(J + mr^2 / G^2\right)$; $\alpha_2 = mg\rho AC_d (r/G)^4 / \left(J + mr^2 / G^2\right)^2$;

$\alpha_3 = B(r/G)^3 / \left(J + mr^2 / G^2\right)^2$; $\alpha_4 = \rho AC_d (r/G)^6 / \left(J + mr^2 / G^2\right)^2$;

$\alpha_5 = 2L_{af} / \left(J + mr^2 / G^2\right) / \left(L_a + L_f\right)$; $\alpha_6 = 2 / (L_a + L_f)$; $\alpha_7 = \alpha_5 \frac{1}{2L_{af}} \rho AC_d \frac{r^3}{G^3}$;

$\alpha_8 = \alpha_5 \frac{1}{L_{af}} \rho AC_d \frac{r^3}{G^3} \omega_0 + \alpha_5 \frac{B}{L_{af}}$.

Fig. 9 shows the robustness test result of a controller designed for a light-weighted EV, compared against a regular PID controller, with parameters as specified in Table 1. To facilitate the graphical representation, the sudden change in hill climbing angle (0.1 rad) is

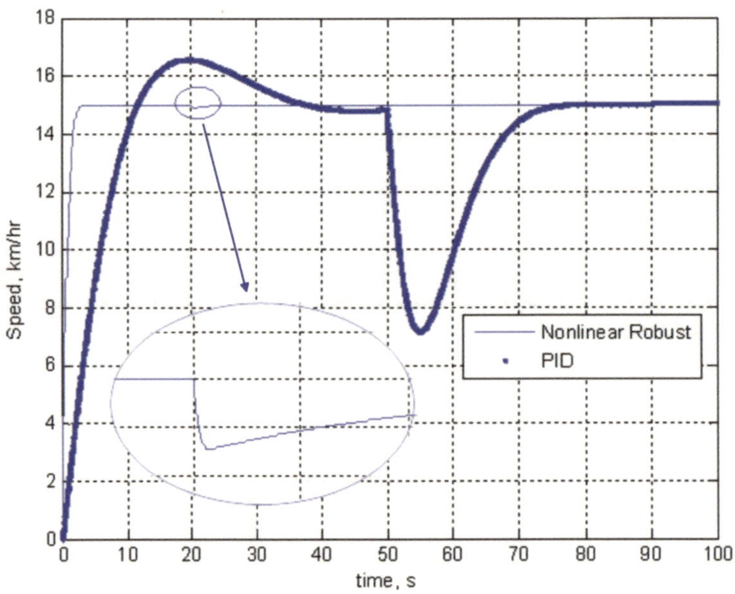


Fig. 9. Robustness test of the non-linear robust controller

applied on the nonlinear robust controller system at t=20 s and on the double-loop PID controller at t=50 s. Shown in Fig. 10 is the excellent robust performance of the designed controller under parameter uncertainties. The test is performed with the arbitrary combinations (i.e., uncertainty in single parameter, two parameters and three parameters) of ±10% uncertainty in the aforementioned three uncertain parameters. Again, for comparison purpose, the performance of (double loop) PID controller and the nonlinear optimal controller are plotted in Fig. 11 and Fig. 12 respectively.

To systematically test the performance, the New European Driving Cycle (NEDC) is used. The New European Driving Cycle is a driving cycle consisting of four repeated ECE-15 driving cycles and an Extra-Urban driving cycle, or EUDC (wry, 2003). The test results are shown in F.g. 13 (the maximum speed is scaled to 50km/h here). It is shown that the nonlinear controller has much better tracking performance than the double loop PID controller, especially in the range of speed below designed nominal speed. And it does not increase much amp-hour consumption (nonlinear optimal: 4.48km/11.97AH; nonlinear robust: 4.825km/10.78AH; PID: 4.49km/10.67AH).

| | Motor | | Vehicle |
|---|---|---|---|
| $L_a + L_f$ (mH) | 6.008 | m (kg) | 800 |
| $R_a + R_f$ (Ω) | 0.12 | A (m²) | 1.8 |
| B (N.M.s) | 0.0002 | $\rho$ (kg/m³) | 1.25 |
| J (kg. m²) | 0.05 | $C_d$ | 0.3 |
| $L_{af}$ (mH) | 1.766 | $\phi$ (°) | 0 |
| $V$ (V) | 0~48 | $\mu_{rr}$ | 0.015 |
| $i$ (A) | 78A (250max) | $r$ (m) | 0.25 |
| $\omega_{nom}$ (r/min) | 2800 ($v$=25km/h) | G | 11 |

Table 1. Parameters of the electric vehicle system

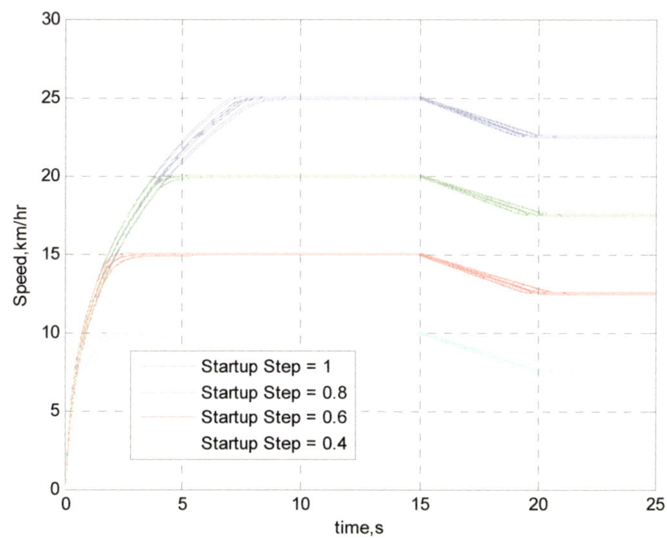


Fig. 10. Robustness test of the non-linear robust controller

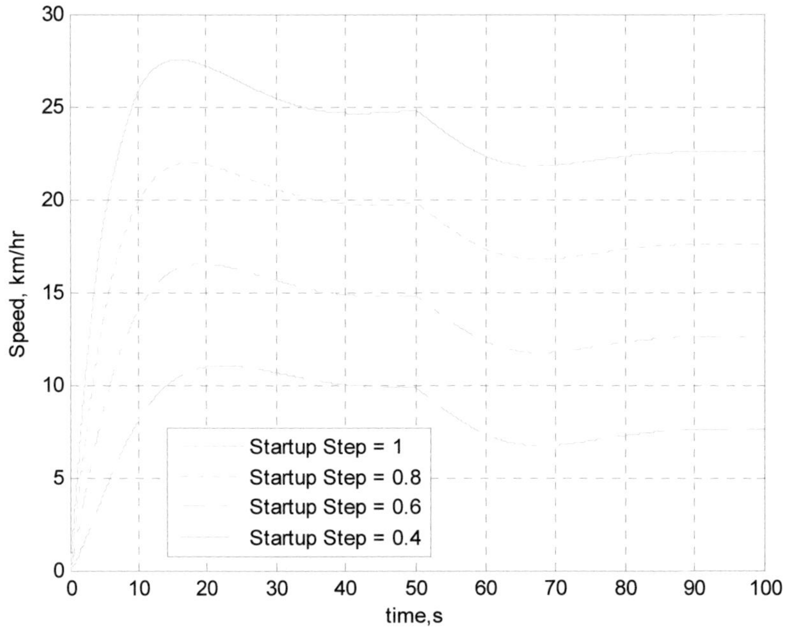


Fig. 11. Performance of non-linear optimal controller

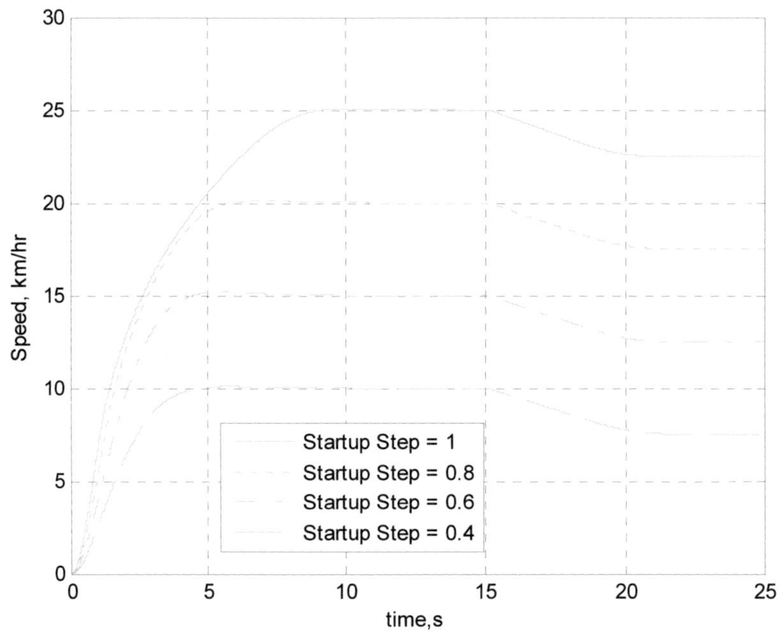


Fig. 12. Performance of double-loop PID controller

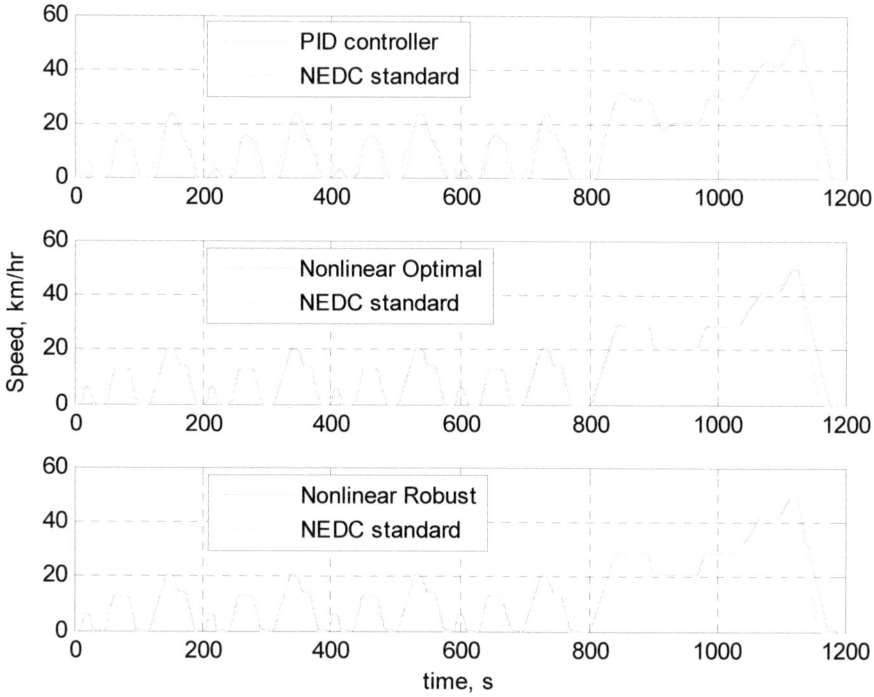


Fig. 13. Results of new European Driving Cycle test

### 4.2 Driven by brushless DC motor

BLDC motor is a closed loop system in nature. The back EMF introduces a negative feedback signal proportional to the motor speed, which enhances the damping of the system. Assume all the initial conditions are zero, the Laplace transform of (7) is:

$$\begin{cases} u_d(s) = E(s) + i(s) \cdot R + L \cdot s \cdot i(s) \\ E(s) = K_e \cdot n(s) \\ T_e(s) = K_T \cdot i(s) \\ T_e(s) = T_L(s) + f\omega(s) + J \cdot s \cdot \omega(s) \end{cases} \tag{21}$$

Therefore, the dynamic model of an EV driven by BLDC can be obtained as shown in Fig. 14.

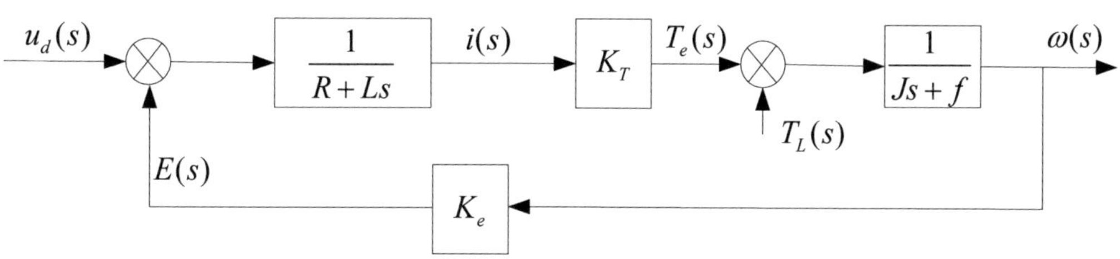


Fig. 14. Dynamic model of an EV driven by BLDC

The transfer function of the inverter can be given as:

$$G_{pwm}(s) = \frac{K_{pwm}}{T_s \cdot s + 1} \quad (22)$$

Where, $K_{pwm}$: gain of inverter; $T_S$: time constant of PWM controller.

The controller of a BLDC is generally composed of current regulation loop and speed regulation loop. In practical systems, due to the fact that the electromagnetic time constant is smaller than electromechanical time constant, current regulation is faster than speed regulation. Hence, speed regulation is faster than the variation of back EMF. Therefore, the effect of back EMF on current regulation loop can be neglected. In order to have small overshoot and good tracking performance, current regulation should be designed as type-I system. Since there are two inertia elements, the current regulator should be designed as PI regulator, whose transfer function is:

$$G_i(s) = K_{pi} \cdot \frac{T_i \cdot s + 1}{T_i \cdot s} = K_{pi} \cdot (1 + \frac{1}{T_i \cdot s}) \quad (23)$$

Where, $K_{pi}$: the proportional coefficient of current regulator; $T_i$: time constant of current regulator.

The structure of the current regulator is shown as the internal loop in Fig. 15. Negnecting the effect of back EMF on the current regulation loop, the stator circuit of the motor can be approximated as a first-order inertia element, hence:

$$\frac{i_s}{u_s} = \frac{1}{R + L \cdot s} = \frac{1/R}{T_a \cdot s + 1} \quad (24)$$

Where, $T_a = L/R$.

The structure of the speed regulation is shown as the external loop in Fig. 15 (Wu et al., 2005). The speed regulation system should have no steady error at steady state and good anti-disturbance capability at transient state, the speed regulator should be designed as type-II system. The speed regulation loop is composed of an integration element and an inertia element, a PI regulator should be used, leading to a transfer function:

$$G_\omega(s) = K_{p\omega} \cdot \frac{T_\omega \cdot s + 1}{T_\omega \cdot s} \quad (25)$$

Where, $K_{pw}$: proportional coefficient of speed regulator; $T_w$: time constant of speed regulator.

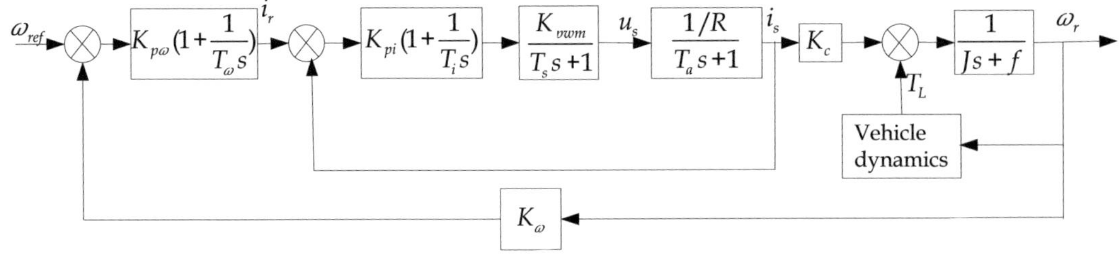


Fig. 15. Structure of the controller for EV driven by BLDC

### 4.3 Driven by PMSM

When designing a controller for PMSM motor, generally two control strategies, i.e., vector control and direct torque control (DTC), are used. In DTC, Stator flux linkage is estimated by integrating the stator voltages. Torque is estimated as a cross product of estimated stator flux linkage vector and measured motor current vector. The estimated flux magnitude and torque are then compared with their reference values. If either the estimated flux or torque deviates from the reference more than allowed tolerance, the transistors of the variable frequency drive are turned off and on in such a way that the flux and torque will return in their tolerance bands as fast as possible. Thus direct torque control is one form of the hysteresis or bang-bang control. In vector control, the stator phase currents are measured and converted into a corresponding complex (space) vector. This current vector is then transformed to a coordinate system rotating with the rotor of the machine. The position can then be obtained by integrating the speed. Then the rotor flux linkage vector is estimated by multiplying the stator current vector with magnetizing inductance $L_m$ and low-pass filtering the result with the rotor no-load time constant $L_r/R_r$, that is, the ratio of the rotor inductance to rotor resistance. Using this rotor flux linkage vector the stator current vector is further transformed into a coordinate system where the real x-axis is aligned with the rotor flux linkage vector. The real x-axis component of the stator current vector in this rotor flux oriented coordinate system can be used to control the rotor flux linkage and the imaginary y-axis component can be used to control the motor torque.

In DTC, the switching speed is low, and the controlled motor generally has low inductance. Therefore, significant current and torque ripples are observed at low speed, limiting the speed regulation range of the controlled system. While the vector control can handle such problems very well (Liu et al., 2004). No matter under low-speed or high-speed condition, the motor current can respond very well once the current waveform required for certain rotation speed is given. The q-axis component of the current is proportional to the torque component needed, giving an excellent performance in dynamic response. Although the vector control algorithm is more complicated than the DTC, the algorithm is not needed to be calculated as frequently as the DTC algorithm. Also the current sensors need not be the best in the market. Thus the cost of the processor and other control hardware is lower making it suitable for applications where the ultimate performance of DTC is not required (Telford et al., 2000). Therefore, vector control is selected in the EV control.

When the number of poles is fixed, the torque of a PMSM is determined by the stator current, therefore, the control of the motor is to control the current. Generally, the stator current $i_s$ is first decomposed in the d-q cordinate, in which d-axis is aligned with the rotor flux and the q-axis is vertical to d-axis.

$$i_s = i_d + ji_q \tag{26}$$

After such decomposition, $i_s$ can be represented as shown in Fig. 16.

It is shown that the stator current is decomposed along the rotor axis $i_d$ component and vertical to the rotor axis $i_q$ component. Then the control of the motor become the calculation of the instantaneous rotor position, i.e., the $\varepsilon$ as shown in Fig. 16. the parameter $\varepsilon$ can be obtained through sensor. Using an intermediate coordinate system $\alpha$, $\beta$ and its phase current projections $i_\alpha$ and $i_\beta$, and the fact that $i_a + i_b + i_c = 0$, one obtains:

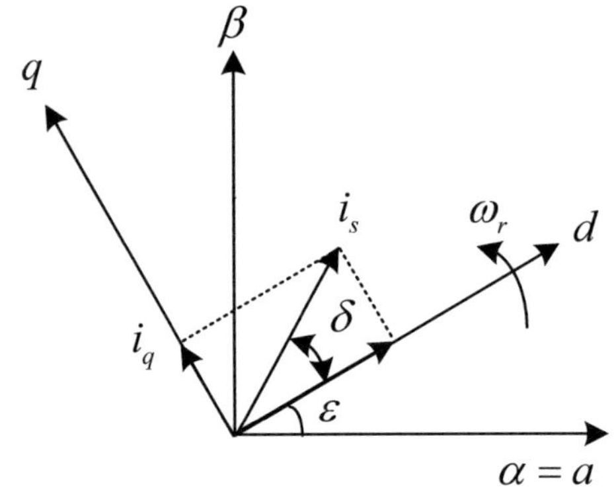


Fig. 16. Current vector decomposition

$$\begin{cases} i_\alpha = i_a \\ i_\beta = \frac{1}{\sqrt{3}} i_\alpha + \frac{2}{\sqrt{3}} i_\beta \end{cases} \tag{27}$$

$i_d$ and $i_q$ are then deduced from $i_\alpha$ and $i_\beta$ by a rotation of the angle ε:

$$\begin{bmatrix} i_d \\ i_q \end{bmatrix} = \begin{bmatrix} \cos\varepsilon & \sin\varepsilon \\ \sin\varepsilon & \cos\varepsilon \end{bmatrix} \begin{bmatrix} i_\alpha \\ i_\beta \end{bmatrix} \tag{28}$$

With proper coordinate transformation, the electric torque of a synchronous motor is described as follows:

$$T_d = P * \Phi * i_q \tag{29}$$

Where, $\Phi$ : flux linkage.

It is shown that, for a magnetically isotropous machine the motor torque depends only on the quadrature $q$ current component (torque component). When $\delta = \pi / 2$ (i.e., $i_d$ = 0), the optimal mode of operation is achieved, where the motor produces the maximum torque.

Currently, vector control (or FOC) is an effective method in variable frequency speed regulation system in synchronous motors. It can obtain large instantaneous speed regulation range. The stator current is measured, then transformed to rotor coordinate with coordinate transformation, forming a current feedback, which can dynamically follow the variation of the current. The speed and rotor position are obtained by optical encoder, forming the speed loop, and finally implement the control of current and speed, greatly enchancing the stability, response speed and control accuracy of the system.

The key to control of PMSM is the realization of high-performance control of motor instantaneous torque. The requirements of torque control for PMSM can be summarized as fast response, high accuracy, low ripple, high efficiency and high power factor, etc. There are three regular methods in current control (Liu et al., 2009): 1) $i_d$ = 0 control; 2) maximum torque and current control; 3) $i_d$ < 0 field weakening control. Other methods such as constant flux linkage control or torque linear control are too complex to use.

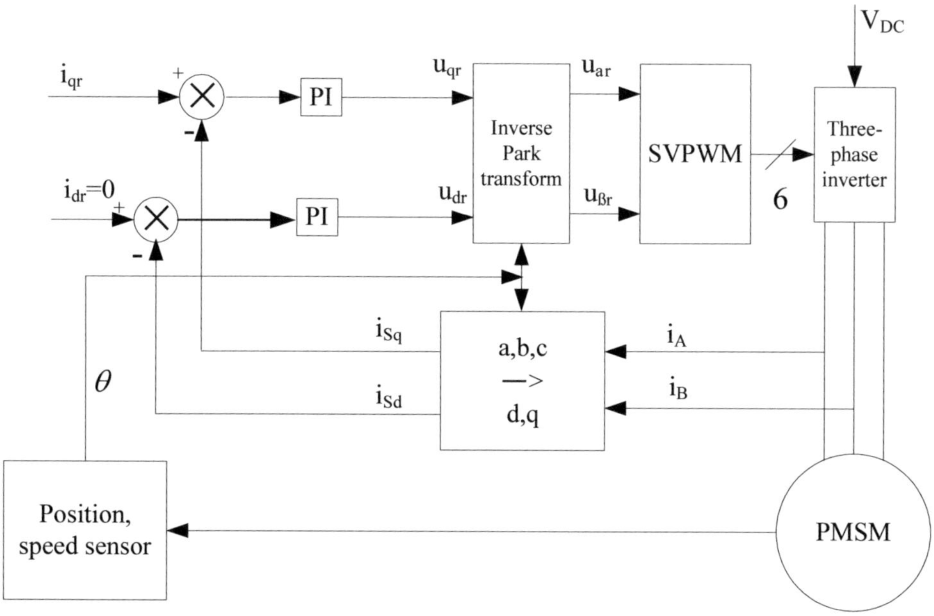


Fig. 17. Vector control of PMSM

$i_d$ = 0 control is simple to implement, and the EM torque and the stator current have a linear relationship, hence having no demagnetizing effect. The disadvantages of this method is the high stator voltage and low power factor at large load conditions, hence cannot fully utilize capacity. In EV control, generally the capacity required is not very large, but it needs high overload capability and good torque response performance. Hence, $i_d$ = 0 control, together with maximum torque-current ratio method, is selected as the solution in EV driven by PMSM. Under such conditions, the structure of the controller is designed as shown in Fig. 17 (Cao & Fan, 2009).

Optical encoder is installed the motor rotor, which accepts the pulse signal when the rotor is rotating. With the pulse signal, the space position $\theta$ and rotor speed $\omega_r$ can be calculated. Using the transforming factor $e^{-j\theta_r}$, $i_d$ and $i_q$ are obtained by performing coordinate transform from ABC-coordinate to dq-coordinate on $i_A$, $i_B$ and $i_C$ (obtained by assuming $i_C = - i_A - i_B$ here). The detected $i_d$ and $i_q$ are compared with the reference value $i_{dr}$ and $i_{qr}$. The error signal is used for regulation, e.g. PI regulator. The inverse Park transform is applied on the calcuated input and then generator output by looking up the table. The generated output is supplied to the three-phase inverter through SVPWM (Space Vector PWM), which controls the motor.

In the SVPWM modulation system, the gain of the inverter can be expressed as:

$$K_{VSI} = \frac{U_0}{2U_\Delta} \tag{30}$$

Where, $U_0$: DC voltage input to the inverter; $U_\Delta$: amplitude of the triangular modulation signal.

The control object of current loop is the DC-AC inverter and the stator circuit of the PMSM motor. DC-AC inverter can generally be regarded as a first-order inertia element with time constant $T_{PWM}$ ($T_{PWM} = \frac{1}{2f_\Delta}$, where $f_\Delta$: frequency of the triangular modulation signal). The transfer function is:

$$G_{VSI}(s) = \frac{K_{VSI}}{T_{PWM}s + 1} \tag{31}$$

With similar discussion in BLDC motor, the stator circuit of the motor is approximated by a first-order inertia element. Introducing a PI regulator, the system is tuned to be type-I element. The PI regulators used for d-axis and q-axis are the same, and the transfer function is:

$$PI_i = K_{Pi}(1 + \frac{1}{T_{ii}s}) \tag{32}$$

In this case, the structure of the current regulation system is shown as the internal loop in Fig. 18. The close loop transfer function of the current regulation system can be simplified as a first-order inertia element:

$$G_{ci}(s) = \frac{G_{oi}(s)}{1 + G_{oi}(s)} = \frac{K_i}{s + k_i} = \frac{1}{sK_i + 1} \tag{33}$$

Where, $K_i = \frac{K_{pi}K_{VSI}}{T_a R_s} = \omega_c$ ; $\omega_c$ : close-loop bandwidth of current loop.

Hence, if a PI regulator is used for speed loop, with transfer function:

$$G_{as}(s) = K_s(1 + \frac{1}{T_s s}) \tag{34}$$

Where, $K_s$, $T_s$: amplification coefficient and time constant of PI regulator respectively.
Then the overall model of control system is shown in Fig. 18, with the outer loop being the speed loop (Brandstetter et al., 2008).

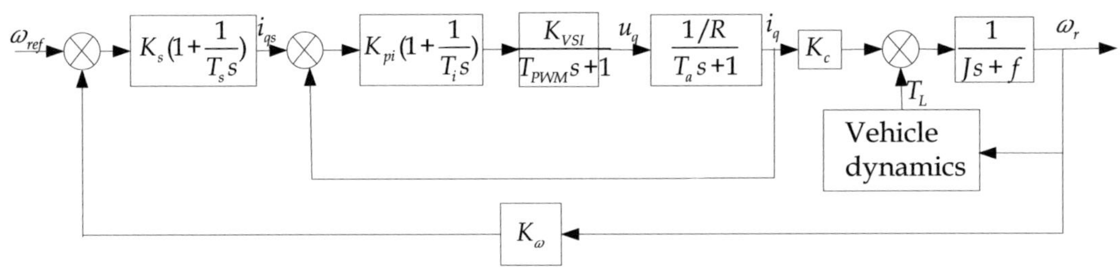


Fig. 18. Control system model of EV driven by PMSM

### 4.4 Driven by inductor motor

Generally, VVVF based on steady motor model, DTC and vector control are used for the speed regulation for induction motor applications. The first method does not consider the complex dynamics inside the electric machine, therefore cannot achieve good dynamic performance. With DTC, although it is not affected by the motor parameters, over-current may occur due to the fact that there is no current feedback. Under low speed condition, the stator flux linkage is circular, and the current is close to sinusoidal. However, at high speed condition, the current waveform is irregular. Large harmonic currents and electromagnetic noises can be observed. Vector control, completely solved above issues. With vector control, the control of AC induction motors is as simple as that of DC motor (Sun & Li, 2002).

The basic principle of vector control is that the magnetic force and power are invariant under normal transform. First transform the model under A-B-C coordinate to $\alpha-\beta$ steady coordinate (Clarke transform), then apply Park transform to the model under $\alpha-\beta$ steady coordinate to d-q rotating coordinate.

Assume the windings of three phase A, B and C are symmetrical, and flowing through balanced sinusoidal current. Then the resultant magnetic force F is rotating with synchronous speed $\varpi_l$.

Assume balanced three phase, the projection that modifies the three phase system into the (α,β) two dimension orthogonal system is (refer to Fig. 16):

$$\begin{cases} i_\alpha = i_a \\ i_\beta = \dfrac{1}{\sqrt{3}} i_a + \dfrac{2}{\sqrt{3}} i_b \end{cases} \tag{35}$$

The (α,β)→(d,q) projection (Park transformation) is the most important transformation in the vector control. This projection modifies a two phase orthogonal system (α,β) in the d,q rotating reference frame. If we consider the d axis aligned with the rotor flux, for the current vector, the relationship from the two reference frame:

$$\begin{bmatrix} i_d \\ i_q \end{bmatrix} = C_{2s/2r} \begin{bmatrix} i_\alpha \\ i_\beta \end{bmatrix} = \begin{bmatrix} \cos\theta & \sin\theta \\ -\sin\theta & \cos\theta \end{bmatrix} \begin{bmatrix} i_\alpha \\ i_\beta \end{bmatrix} \tag{36}$$

Whose inverse transformation is:

$$\begin{bmatrix} i_\alpha \\ i_\beta \end{bmatrix} = \begin{bmatrix} \cos\theta & -\sin\theta \\ \sin\theta & \cos\theta \end{bmatrix} \begin{bmatrix} i_d \\ i_q \end{bmatrix} \tag{37}$$

The model of a three-phase induction model under d-q two-axis coordinate can be expressed as follows.

Votlage equations:

$$\begin{bmatrix} u_{ds} \\ u_{qs} \\ u_{dr} \\ u_{qr} \end{bmatrix} = \begin{bmatrix} r_s + L_s p & -\varpi_l L_s & L_m p & -\varpi_l L_m \\ 0 & r_s + L_s p & \varpi_l L_s & L_m p \\ L_m p & -\varpi_r L_m & r_r + L_r p & \varpi_s L_r \\ \varpi_s L_m & L_m p & \varpi_s L_r & r_r + L_r p \end{bmatrix} \cdot \begin{bmatrix} i_{ds} \\ i_{qs} \\ i_{dr} \\ i_{qr} \end{bmatrix} \tag{38}$$

Where, $L_s = L_{ls} + \frac{3}{2} L_{ms}$, $L_r = L_{lr} + \frac{3}{2} L_{ms}$, $L_m = \frac{3}{2} L_{ms}$, $p = \frac{d}{dt}$; $\varpi_s = \varpi_l - \varpi_r$, $\varpi_l$: synchronous rotating speed.

Flux linkage equations:

$$\begin{bmatrix} \psi_{ds} \\ \psi_{qs} \\ \psi_{dr} \\ \psi_{qr} \end{bmatrix} = \begin{bmatrix} r_s & 0 & L_m & 0 \\ 0 & r_s & 0 & L_m \\ L_m & 0 & L_r & 0 \\ 0 & L_m & 0 & L_r \end{bmatrix} \cdot \begin{bmatrix} i_{ds} \\ i_{qs} \\ i_{dr} \\ i_{qr} \end{bmatrix} \tag{39}$$

Torque equation:

$$T_e = PL_m(i_{qs}i_{dr} - i_{ds}i_{qr}) \tag{40}$$

With these transformations, the stator current is decomposed into two DC components aligned with rotor magnetic field, $i_d$ and $i_q$, with $i_d$ corresponding to excitation current and $i_q$ corresponding to torque current. These two currents can be controlled separately. Control of $i_d$ means control the flux, while control of $i_q$ means control of torque, which is like the control of a DC motor, and then different control strategies can be easily applied on the control of EV driven by induction motor. With vector control scheme, the structure of whole system can be given as shown in Fig. 19.

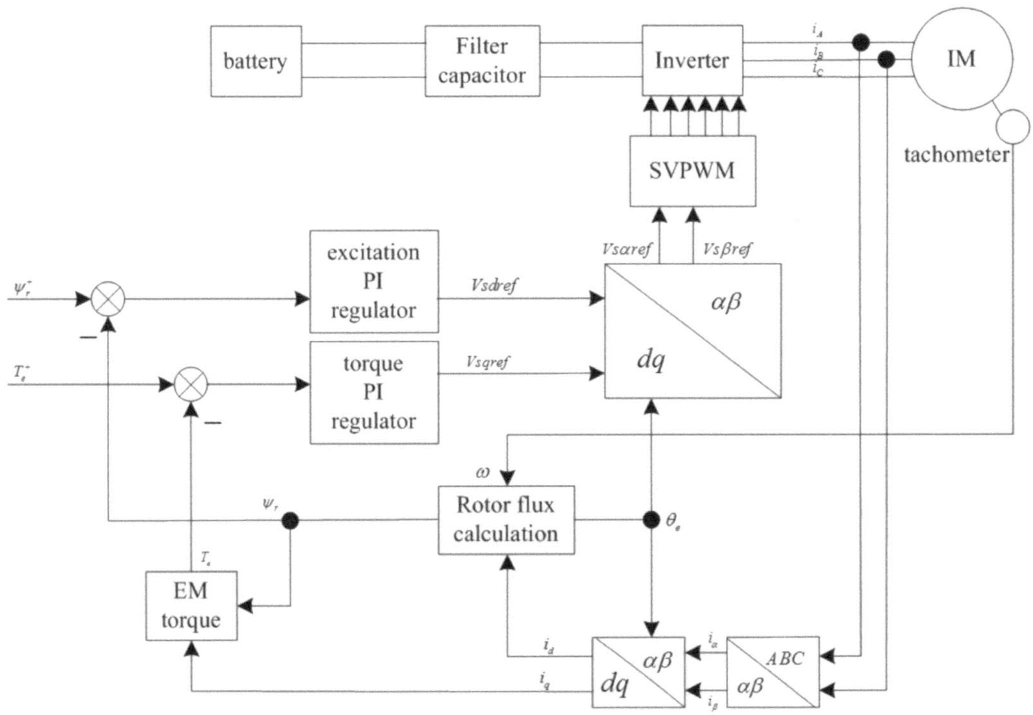


Fig. 19. The control system of EV driven by induction motor with vector control scheme

## 5. Implementation of electric vehicle controller with DSP

As the development of the high computing capability microprocessor, such as DSP (Digital Signal Processor), it is possible to perform advanced control strategies on the electric vehicle and integrate complex functions to achieve optimal performance (Huang et al., 2007). These capabilities can be utilized to enhance the performance and safety of individual vehicles as

well as to operate vehicles in formations for specific purposes (Huang et al., 2007). In this section, the design and implementation of an EV controller with DSP will be discussed.
An EV controller may preferably have following functions:

- Due to the frequent acceleration/deceleration and up/down of EV, it is preferably that the regenerative control function is included, to save energy in battery and hence enhance the driving range per charge;
- Self diagnostics and restoration from failure;
- Scalable to function expansion;
- And comprehensive protection functions, including over-temperature protection, over-voltage and under-voltage protection, over-current protection, short-circuit protection, motor-lock protection, and protection for control unit, main switch and security, etc.

Fig. 20 shows detailed implementation of a EV controller with DSP (Huang et al., 2007). The whole system is composed of power circuit and control circuit. The control circuit include two parts. One is the external control circuit including auxiliary power supply, MOSFET driving circuit, isolation & protection circuit and contactor driving circuit. These circuits have common ground with the battery group. Another one is the kernel control circuit, including DSP kernel TMS320F2812, variant protection circuits, detection circuit, signal conditioning circuit, CAN communication circuit, external watchdog circuit and isolated auxiliary power system, etc.

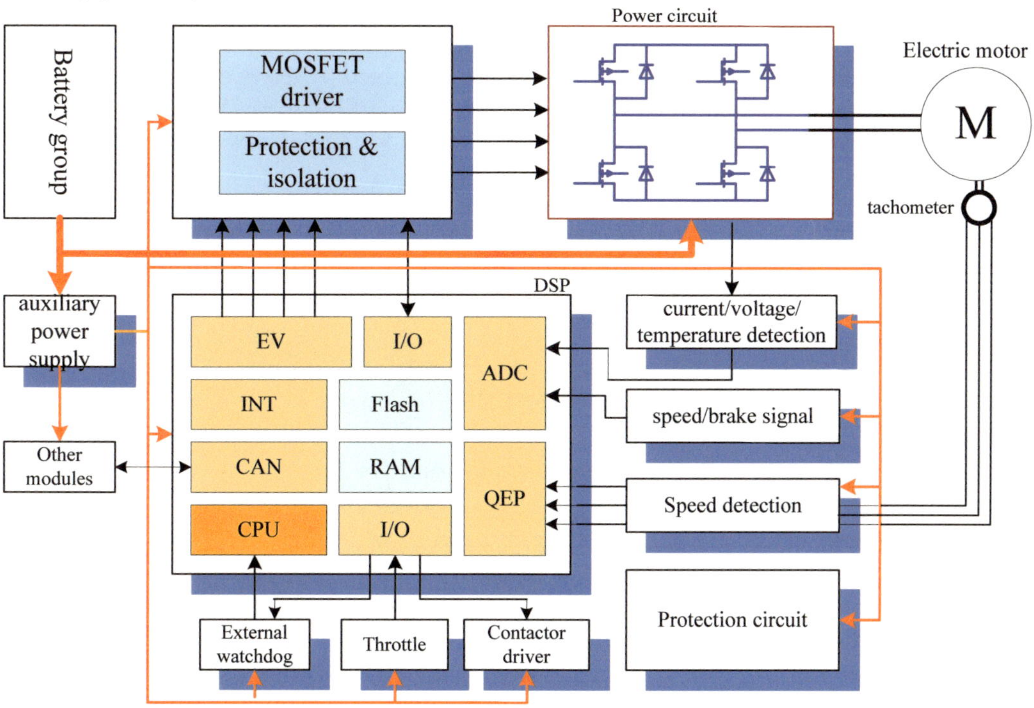


Fig. 20. Structure of a DSP-based EV controller

The control flow of the whole system, as shown in Fig. 20, is that the DSP generate PWM modulation signal, which is used to drive the power converter (chopper for DC motor and inverter for AC motor) after isolation and amplification. The power converter controls output of the battery to control the speed of the motor, and implement the energy regenerative control. The voltage, current and speed signal of motor are detected and sent to DSP after signal conditioning, to form a feedback control.

The control circuit detects the voltage, current, temperature and rotation speed to realize the function of regulation and protection, and provides support for implementation of advanced control strategies. The voltage, current, and temperature are measured through the quantification of AD converter in DSP kernel. The speed is detected with the QEP (quadrature encoder pulse) inside the DSP, and the speed signal is used for realization of loss-of-control protection, smooth startup control, anti-skip control during up-hill startup and providing speed information for the driver. Speed regulation (throttle) and breaking signal are generated by proper device and acquisited by AD of the DSP after signal conditioning.

The gear shift is digital signal. It is input to the DSP through I/O after isolation and amplitude limiting. The main contactor is connected in series with the power circuit. The winding of the contactor is controled by digital signal output from the I/O of DSP, to switch on or off the power circuit. In order to restore the whole system from severe failure, double watchdog (internal and external) is used. When the DSP is invalidated, the external watchdog can reset the chip to restore the normal operation of the controller. Even the DSP chip is destroyed, the output of the controller is switched off due to the continuing resetting. With this scheme, the security of the EV is assured even under severe failure in the controller.

The power converter is controlled by PWM, which is generated by two event managers (with one generating 16 PWM signals). The PWM's generated by the event managers of DSP are sent to drive the power converter after isolation and amplification. When there are faults in the power converter or in driving circuit, the fault information is sent to DSP through its I/O after isolation.

The configuration of controller parameters, various state information and fault information are exchanged with external devices, such as portable computer, handheld programable devices, or display monitor on the EV, through CAN (controller area network) bus, which is controlled by the embedded CAN controller in DSP.

The power converter is composed of semiconductor power electronics devices and their snubber circuit and freewheeling diodes. Power semiconductor is the key for power conversion in EV controller. Now, new generation of power semiconductor devices are developed and the performances are improved continuously. Presently, in most of EV control applications, MOSFET (metal–oxide–semiconductor field-effect transistor) or IGBT (insulated-gate field-effect transistor) is used. IGBT needs higher driving power, but has lower working frequency. MOSFET has the advantage of simple driving circuit and low conducting resistance, making it specially suitable for driving application in high-current low-voltage motor. In this chapter, the results are all based on MOSFET.

The main flow chart of the software running inside the DSP processor is shown in Fig. 21. It is a state-based processing system, in which the status of the whole EV system is detected in a real-time manner, various functions are activated once certain failure is detected or specific order is received. The maximum delay of processing different failure is one main-loop processing cycle.

Fig. 22 is the test platform in laboratory, in which the load torque is a brake dynamometer. Shown in Fig. 23 and Fig. 24 are the performance of regenerative braking and speed regulation with a PMSM motor.

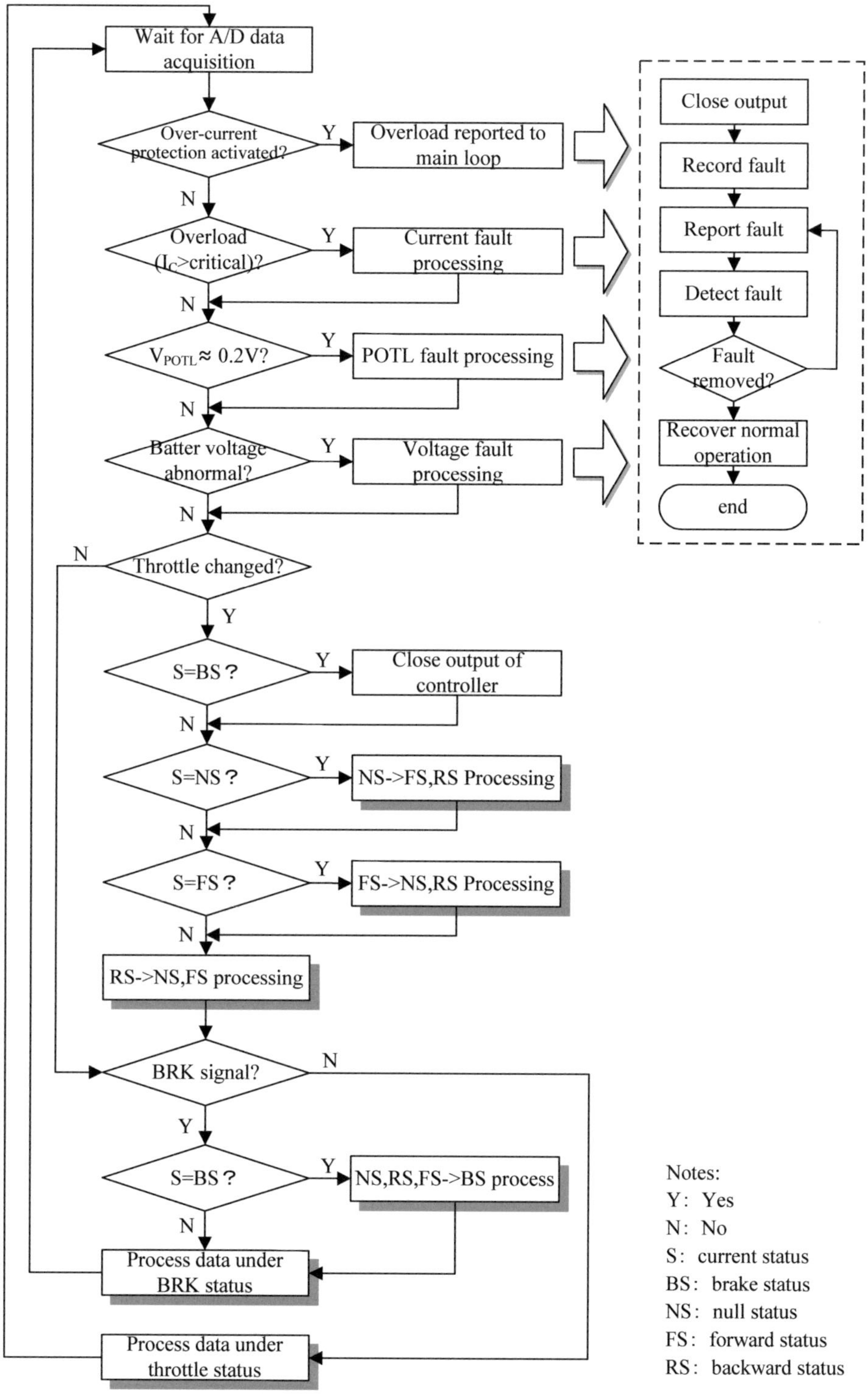


Fig. 21. Operation of the EV controller

Fig. 22. Test platform of controller

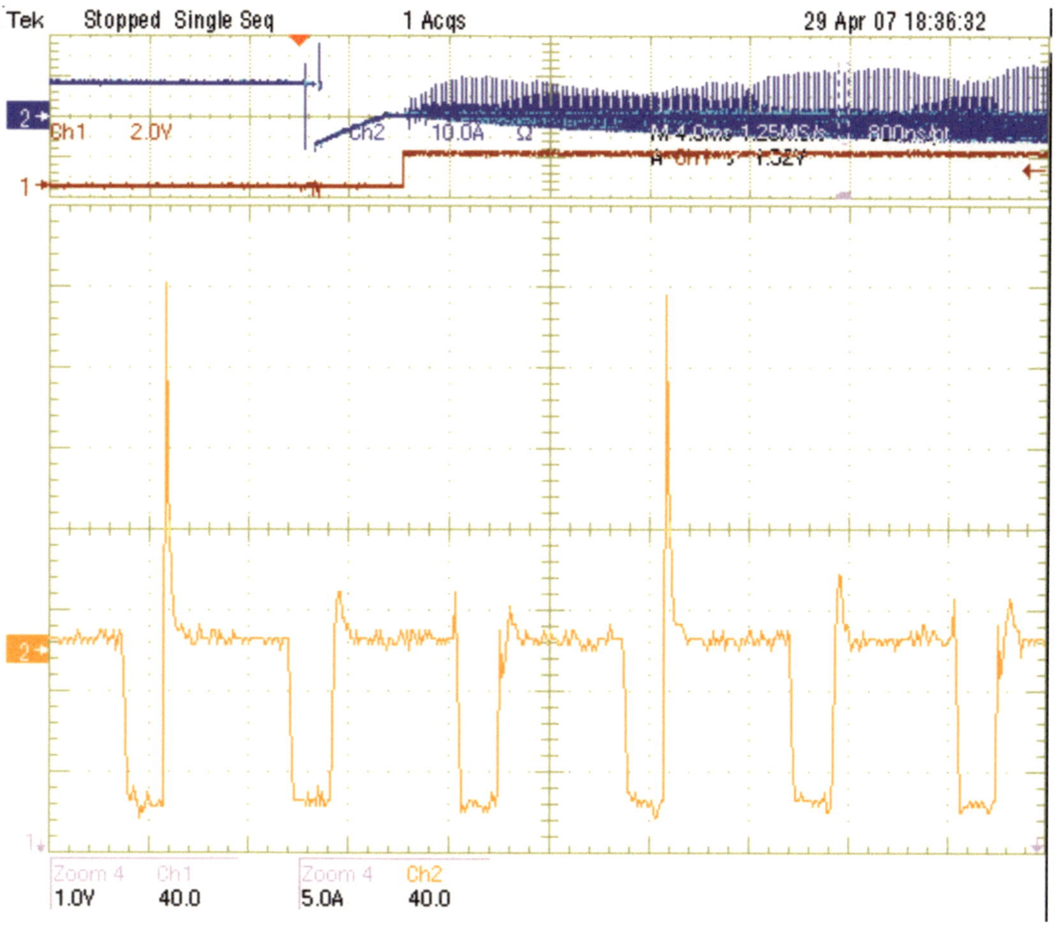


Fig. 23. Waveform of battery charging current under regenerative braking

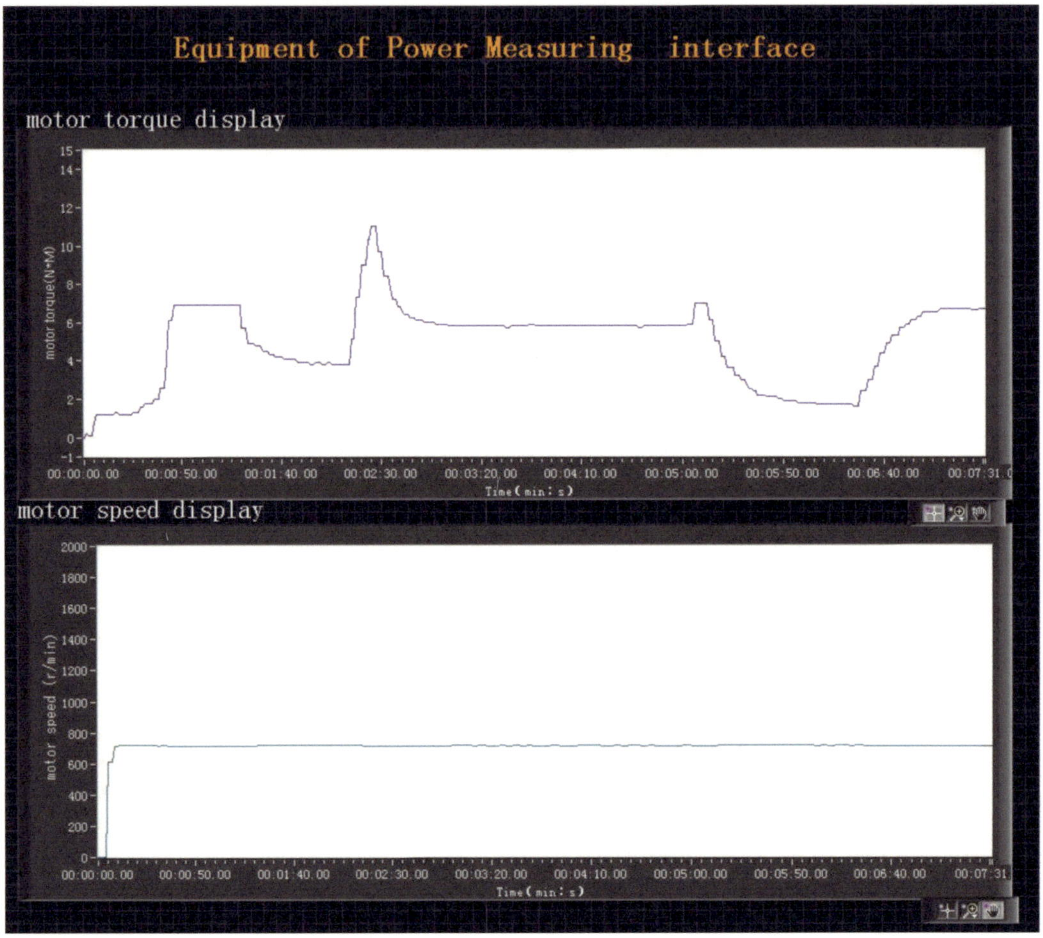


Fig. 24. Performance test of speed control of the designed controller

## 6. Conclusion

In this chapter, the modeling of electric vehicle is discussed in detail. Then, the control of electric vehicle driven by different motors is discussed. Both brushed and brushless DC (Direct Current) motors are discussed. And for AC (Alternative Current) motors, the discussion is focused on induction motor and permanent magnet synchronous motor. The design of controllers for different motor-driven electric vehicle is discussed in-depth, and the tested high-performance control strategies for different motors are presented. The model-based controller is designed for brushed DC motor, while for other motors, model based controllers can be designed in a similar way if the control strategies discussed in this chapter are used. Finally, the implementation of the controller with DSP and some test results with this platform are presented.

## 7. References

wry, J.(2003). *Electric Vehicle Technology Explained.* pp. 183 - 195, John Wiley & Sons, Ltd., ISBN 0-470-85163-5, UK

Cheng, Y., Van Mierlo, J., Van den Bossche, P.; Lataire, P. (2006). Energy Sources Control and Management in Hybrid Electric Vehicles. *Proc. 12th Int. Power Electronics and Motion Control Conf.*, pp. 524 - 530, Portoroz, Slovenia, Aug. 2006

Sakai, S., Hori, Y. (2000). Advanced Vehicle Motion Control of Electric Vehicle Based on the Fast Motor Torque Response. 5th *International Symposium on Advanced Vehicle Control*, pp. 1-8 Michigan, USA, Aug., 2000

Liu, Q., Zhong, Y., Zhou Z. (2004). Research of drive control system in electric vehicle based on DSP. *Proc. 7th Int. Conf. Signal Processing*, pp. 539 - 542, Beijing, China, Aug. 2004

Lin, J., Kanellakopoulos, I. (1995). Nonlinear Design of Active Suspensions. *Proc. 34th IEEE Conf. Decision and Control*, pp.1-3, New Orleans, United States, Dec. 1995

Poorani, S., Kumar, K.U., Renganarayanan S. (2003). Intelligent controller design for electric vehicle. *Proc. 57th IEEE Semiannual Vehicular Technology Conf.*, pp. 2447 - 2450, Jeju, Korea, Apr. 2003

Khatun, P., Bingham, C.M., Schofield, N., Mellor, P.H. (2003). Application of fuzzy control algorithms for electric vehicle antilock braking/traction control systems. *IEEE Trans. Vehicular Tech.*, Vol.52 (5), 2003, pp. 1356 - 1364

Profumo, F., Madlena, M., Griva. G. (1996). State variables controller design for vibrations suppression in electric vehicles. *Proc. 27th Annual IEEE Power Electronics Specialists Conference*, pp. 1940 - 1947, 1996

Chan, C.C. (1999). The Past Present and Future of Electric Vehicle Development. *IEEE Power Electronics and Drive Systems*,1999, pp.11-13

Mehta, S., Chiasson, J. (1998). Nonlinear control of a series DC motor: theory and experiment. *IEEE Transactions on industrial Electronics*, Vol. 45 (1), 1998, pp. 134 - 141

Wu, H.X., Cheng, S.K, Cui, S.M. (2005). A Controller of Brushless DC Motor for Electric Vehicle. *IEEE Transactions on Magnetics*, Vol. 41 (1), 2005, pp. 509-513

Cui, B.W., Zhou, J.H., Ren Z. (2001). Modeling and simulation of permanent magnet synchronous motor drives. *Proceedings of the Fifth International Conference on Electrical Machines and Systems*, pp. 905 - 908, 2001

Gosden, D.F., Chalmers, B.J., Musaba, L. (1994). Drive system design for an electric vehicle based on alternative motor types. *IEE Power Electronics and Variable-speed Drives Conference*, pp. 710-715, 1994

Dilmi, S., Yurkovich, S. (2005). Nonlinear Torque Control of the Induction Motor in Hybrid Electric Vehicle Applications. *American Control Conference*, pp. 3001-3006, 2005, Portland, OR, USA

Huang, Q., Huang, Z., Zhou, H.L. (2009). Nonlinear Robust and Optimal Control of a Light-weighted All Electric Vehicle. *IET Control theory & applications*, Vol.3 (4), 2009, pp. 437-444

Marino, R., Tomei, P. (1993). Robust stabilization of feedback linearizable time-varying uncertain nonlinear systems. *Automatica*, Vol.29 (1), 1993, pp. 181-189

Freeman, R. A., Kokotovic, P. V. (1996). Robust Nonlinear Control Design ---- State-Space and Lyapunov Techniques. pp. 107-117, Birkhauser, 1996

Liu, Q., Zhong, Y., Zhou Z. (2004). Research of drive control system in electric vehicle based on DSP. *Proc. 7th Int. Conf. Signal Processing,* pp. 539 - 542, Beijing, China, Aug. 2004

Telford, D., Dunnigan, M.W., Williams, B.W. (2000). A comparison of vector control and direct torque control of an induction machine. *IEEE 31st Annual Power Electronics Specialists Conference*, pp. 421 - 426, 2000

Liu, J., Huang, M.Z., Wang, Y. (2009). Research on Vector-Control System of PMSM Based on Internal Model Control of Current Loop. *Second International Workshop on Computer Science and Engineering,* pp. 297 - 301, 2009

Cao, X.Q., Fan, L.P. (2009). Efficiency-optimized vector control of PMSM drive for hybrid electric vehicle. *International Conference on Mechatronics and Automation,* pp.423 - 427, 2009

Brandstetter, P., Krecek, T., Korbel, P. (2008). Non-model based sensorless vector control of permanent magnet synchronous motor. *IEEE Int. Symp. Ind. Electronics,* pp. 618 - 623, 2008

Sun, J.J., Li, Y.D. (2002). Voltage-oriented vector control of induction motor: principle and performance improvement. Proc. Power Conversion Conference, pp. 1340 - 1345, 2002

Huang, Z. (2007). Design and Implementation of an EV controller. MS thesis, pp. 18-26, 2007, University of Electronic Science and Technology of China, Chengdu, China

Huang, Z., Huang, Qi, Chen, Q. (2007). Design of DC Motor Controller for Electric Vehicle. Vehicle & Power Technology, No. 2, 2007, pp. 52-56, 64